T0096874

RAND

A Guide for Analysis Using Advanced Distributed Simulation (ADS)

Thomas Lucas, Robert Kerchner,
John Friel, Daniel Jones

Prepared for the
United States Air Force

Project AIR FORCE

Approved for public release; distribution unlimited

Preface

The purpose of this report is to assist those in the vanguard of using Advanced Distributed Simulation (ADS) for analysis. The report discusses a broad range of issues critical to successful ADS-supported analyses. Major topic areas include potential ADS analysis strengths and weaknesses, the role ADS might play within a broader analysis strategy, experimental design, exercise preparation and management, and post-exercise analysis. Because it is impossible to comprehensively treat all of these subjects, we emphasize the breadth of analysis issues over depth in their coverage—with references to more detailed resources. Furthermore, the depth of coverage is highly variable. The greatest detail is provided on the roles of ADS in the analysis process and in experimental design.

This work builds on previous efforts that examined how the Air Force might effectively use ADS to enhance its analysis capabilities. The report is written from the perspective of analysts who have spent decades using traditional analysis methods and are trying to determine how best to utilize the new capabilities ADS offers. The need for such a guide is well recognized by the military analyst community, as evidenced by the extensive discussions at the 1996 Military Operations Research Society (MORS) workshop on Advanced Distributed Simulation for Analysis (ADSA) and at the semi-annual Distributed Interactive Simulation (DIS) meetings.

There are three primary audiences we believe might benefit from this research:

- Decisionmakers, such as program managers, who need to determine how ADS might support their analysis needs and how to interpret ADS analysis products.

- Technologists, modelers, and trainers, who compose the majority of the ADS community, who are experienced in ADS but without an extensive analysis background. This report may help this community better understand analytic requirements and what it takes to produce credible analysis products.

- Analysis directors and their staffs, who must design, conduct, and communicate ADS-supported analysis, particularly, those organizations that are new to ADS and can benefit from others' lessons learned.

This study is based on our investigations of several pioneering ADS analysis efforts and the application of back-to-basics scientific principles. We emphasize the analysis process and how ADS fits into it, rather than replacing it. In highlighting special analytic needs when using ADS, it is sometimes necessary to repeat what experienced analysts are already familiar with.

What we detail in this report is an ideal. As a consequence of practical resource constraints, it may not always be feasible to carry out all of the steps we recommend—nor will all study specifics conform to the generalizations necessitated by such a broad scope. Nonetheless, this report can serve as a guide when using ADS or when considering its use.

This work was performed in the Effective Application of Advanced Distributed Simulation project in RAND's Project AIR FORCE. This project is sponsored by Major General Tom Case of AF/XOM. The report should be of interest to analysts, the greater ADS community, and those who might commission or consume ADS analyses in the military departments and the Office of the Secretary of Defense.

Project AIR FORCE

Project AIR FORCE, a division of RAND, is the Air Force federally funded research and development center (FFRDC) for studies and analyses. It provides the Air Force with independent analyses of policy alternatives affecting the development, employment, combat readiness, and support of current and future aerospace forces. Research is performed in three programs: Strategy, Doctrine, and Force Structure; Force Modernization and Employment; and Resource Management and System Acquisition.

Contents

Figures

Table

Summary

Part of the promise of Advanced Distributed Simulation (ADS) and justification for the large investment in ADS technologies is that ADS will revolutionize how analysts do business. Presently, with ADS still in its infancy, the majority of its uses have been technology development, training, and demonstrations; however, ADS for analysis is rapidly becoming a reality and may affect important decisions. Pioneering analysis efforts include the Airborne Laser (ABL) tests, the Anti-Armor Advanced Technical Demonstration (A2ATD) experiments, and a joint Air Force and Navy effort to study whether and how the Cooperative Engagement Capability (CEC) should be extended to the Airborne Warning and Control System (AWACS). These efforts are finding a high potential benefit and a steep learning curve associated with ADS. The intent of this guide is to acquaint potential users and consumers with ADS so that it is used properly, and does not suffer from excessive optimism followed by dashed expectations.

The Analytic Potential of ADS

In theory, there are potentially many analytic benefits from ADS technologies that extend and augment traditional analytic methods. It is ADS technologies that can provide the foundation for building the Joint Synthetic Battlespace (JSB) at the heart of the Air Force's *A New Vector* (1995). The other services and the Department of Defense (DoD) have similar hopes for ADS that stem from the vision that ADS will mature enough to be able to realistically simulate a seamless synthetic theater of war (STOW)—with joint live, virtual, and constructive elements participating. The elements will all share the same virtual battlespace even though they are hosted at distributed homebases. While the JSB ideal is still only a vision, properly used, today's ADS can provide tremendous analytic utility. ADS currently has the capability to

- provide a realistic treatment of human performance, a notable weakness with constructive simulations,

- obtain insights into the cause-and-effect "drivers" of combat, which can be extremely useful when developing tactical concepts to enhance or defend against new weapons systems,

- communicate analytic results to decisionmakers more effectively,

- facilitate multidisciplinary research teams that explicitly include warfighters—thus accruing credibility, and

- enable the combining of multiple disparate service simulations into a single joint simulation for theaterwide scenarios with service-accredited models at previously unobtainable levels of detail.

Challenges in Analytic Applications of ADS

Significant challenges must be overcome before the full ADS analysis potential can be realized. Some of the more important are:

- The sheer *complexity* of a distributed joint STOW. Each component has its own specific assumptions and limitations. Accounting for these is critical in determining whether simulation results are credible or merely simulation artifacts.

- The difficulties associated with *exclusive* use of human-in-the-loop (HITL) analysis. The most important are (1) restriction to real-time, which precludes exploring many scenario variations or achieving statistical precision, and (2) human factors such as getting representative samples, participant learning and gaming, participant boredom, and exact reproducibility.

- The *logistical load and expense* of distributed efforts, which are significantly greater than for single-suite simulations and simulators. Not only are the simulations distributed, but the expertise and much of the data are as well.

ADS Within a Broader Research Plan

The analytically oriented HITL ADS projects we have seen logically require a balanced mix of ADS and more traditional methods, and should not exclusively rely on ADS exercises as the source of analytic information. In fact, we see a natural synergy between ADS and traditional methods; each supplies strength to the other's weaknesses. Traditional methods allow for greater control and more factors to be varied—ideal for identifying critical variables or scenarios. ADS facilitates joint high-resolution scenarios with warfighters representing the important human dimension.

The above suggests the following analytic roles for ADS and traditional analysis: Use ADS primarily to inform about human performance factors in constructive models, cross-check constructive model results, and assess warfighter elements in a few carefully designed scenarios; use (relatively) inexpensive constructive models primarily to focus the limited ADS runs on the most important cases and perform the bulk of the exploration (after being informed by ADS). This constructive-ADS iteration continues as time and money permit.

Of course, there may be issues, such as human factors, for which ADS can reliably be used exclusively. Furthermore, there may be issues that can be effectively addressed by traditional stand-alone constructive methods. For the many issues that benefit from a mixture of both ADS and traditional methods, Figure S.1 illustrates how this concept might look in the context of an analysis where ADS plays a significant role. Although the process shown here is idealized, we believe that it serves both as a practical guide for combined ADS and traditional analysis and as a goal to be achieved. ADS is used in three distinct ways in the scheme. The ADS experiment block (Block 1) in Figure S.1 refers to the use most visible to the consumers of the analysis, and corresponds to the high-value ADS runs for scenarios of interest. A second use of ADS is human-in-the-loop experiments aimed at human performance factors (HPFs) in the constructive simulations (Block 2). Furthermore, ADS can be used in a preliminary *exploratory* manner to identify HPFs that are likely scenario drivers (also within Block 2).

Key ADS Analysis Issues

Some critical points above and beyond the central theme of ADS in a broader analysis perspective include:

- Interoperability is not guaranteed by compliance to standards. Unless great care is taken, the lack of interoperability will bias simulation outcomes. Key aspects to consider are differences in data, algorithms, resolution, terrain, visual displays, and human participants. Often these are difficult to compare theoretically. We recommend interoperability be studied through an iterative series of increasingly larger empirical tests among components.

- HITL ADS runs are a precious commodity and must be designed with great care, rather than executed as free play. The ADS runs will be most valuable if they are designed to address specific hypotheses.

- The high dimensionality and few samples available in HITL ADS experiments mean the effort will benefit from advanced design of experiment (DOE) techniques.

- Analysis in a training environment greatly restricts the types of analysis one can perform.

- A successful effort requires multidisciplinary participation in the total analysis process, to include analysts, site managers, modelers, operators, other warfighters, and network managers.

- The complexity of large distributed efforts puts an added burden on testing and rehearsals. The rehearsals should include a mock analysis to ensure the needed information can be obtained.

- Post-exercise analysis should include after-action reviews, statistics, visual replays, and allow for the insertion of new objects for visualization and analysis.

- Model and data freeze dates must be established and adhered to.

- ADS experiments involve an inevitable reduction in reliability, i.e., simulation or network failure. The situation should be planned for—including a real-time contingency playbook.

Conclusion

We believe that ADS has great potential for increasing the effectiveness, scope, and depth of analysis, but the role of ADS in an analysis must be carefully specified. In combination with traditional methods, ADS can more credibly represent human interactions and improve this critical component of our models, whereas traditional methods can be used to examine a greater breadth of cases and focus on those conditions where ADS methods are essential. These benefits will not be gained without overcoming a variety of technical, operational, and administrative challenges. In particular, we feel that resolving problems with interoperability among models is essential. Unfortunately, "plug and play" interoperability has not been successfully addressed in contexts that are much simpler than distributed combat simulation. Thus, there is little reason to expect that these challenges can be successfully solved, for general ADS combat analysis purposes, in the near future.

To improve model interoperability we need to establish well-accepted approaches to representing combat elements, document the models and standards used, build up trusted and tested implementations through frequent and wide use of the models, and provide easy accessibility to the models. Further research in these areas is needed if ADS is to become an oft-used and credible vehicle for analysis. Moreover, given that we believe ADS is often best used in conjunction with stand-alone constructive simulations, investments must also be made in these models and the analysis methods that use them.

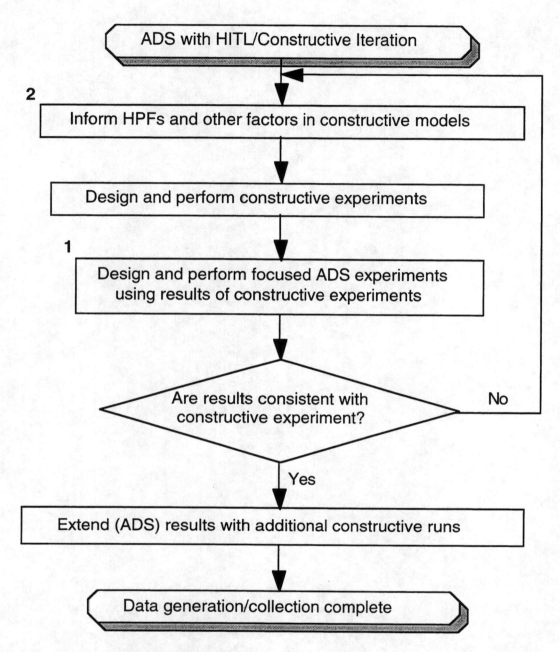

Figure S.1—Interplay of Constructive and ADS/HITL Experiments in an Analysis Effort

Acknowledgments

The foundation of much of this research is observations we made at several ADS exercises, as well as numerous meetings with many of those in the vanguard of ADS analysis efforts. Each contributed to the ideas within this report. At the risk of missing someone, the following people have made particularly important contributions. Colonel Ed Crowder of Air Force Studies and Analyses Agency (AFSAA) has provided valuable feedback to earlier briefings and connected us with several ongoing efforts. Special thanks are due Lieutenant Colonel Denny Lester and his staff at the Theater Air Command and Control Simulation Facility (TACCSF), who hosted several visits and shared many of their hard-earned lessons and insights with us. Additionally, we have benefited immensely by participating in other community efforts looking at ADS for analysis. These include the Analysis and Engineering Tiger Team run by Dr. Dale Pace of Johns Hopkins' Applied Physics Laboratory (APL); the Military Operations Research Society (MORS) ADS for analysis workshop; the Exercise, Management and Feedback (EMF) working group, and the Credible Uses (CU) of Distributed Interactive Simulation (DIS) for Analysis special interest group at the semi-annual DIS workshops. The latter two efforts were run by Mr. William Tucker and LTC Pat Vye, respectively. Finally, reviews by Bart Bennett, Randy Steeb, and Monti Callero of RAND, and Jim Hodges of the University of Minnesota, have improved the clarity and content of the report.

List of Symbols

AAR	After Action Review
A2ATD	Anti-Armor Advanced Technical Demonstration
ABL	Airborne Laser
ADS	Advanced Distributed Simulation
ADSA	Advanced Distributed Simulation for Analysis
A&E	Analysis and Engineering
AFSAA	Air Force Studies and Analyses Agency
AG	Application Gateway
ALSP	Aggregate Level Simulation Protocol
AMSAA	Army Materiel Systems Analysis Activity
APL	Applied Physics Laboratory
AWACS	Airborne Warning and Control System
AWE	Advanced Warfighting Experiment
C4ISR	Command, Control, Communications, Computers, Intelligence, Surveillance, & Reconnaissance
CAGIS	Cartographic Analysis and Geographic Information System
CEC	Cooperative Engagement Capability
CGF	Computer Generated Forces
COA	Course of Action
CU	Credible Uses
DARPA	Defense Advanced Research Projects Agency
DBMS	Database Management System
DIA	Defense Intelligence Agency
DIS	Distributed Interactive Simulation
DMA	Defense Mapping Agency
DoD	Department of Defense
DOE	Design of Experiment
DPA&E	Defense Program Analysis and Evaluation

EMF	Exercise, Management and Feedback
FOM	Federation Object Model
GPS	Global Positioning System
HITL	Human-in-the-loop
HLA	High-Level Architecture
HPFs	Human Performance Factors
IST	Institute for Simulation and Training
JETS	Joint Engagement Technology Study
JSB	Joint Synthetic Battlespace
MADAM	Model to Assess Damage to Armor by Munitions
M&S	Modeling and Simulation
MOE	Measure of Effectiveness
MOM	Measure of Merit
MOP	Measure of Performance
MORS	Military Operations Research Society
NRaD	Naval Command, Control and Ocean Surveillance Center RDT&E Division
ODS	Operation Desert Storm
OMT	Object Model Template
OPFOR	Opposing Force
OSD	Office of the Secretary of Defense
PDU	Protocol Data Units
QDR	Quadrennial Defense Review
RJARS	RAND Jamming and Radar Simulation
RTAM	RAND Target Acquisition Model
SAFOR	Semi-Automated Forces
S&T	Science and Technology
SEMINT	Seamless Model Interface
SOM	Simulation Object Model
STOW	Synthetic Theater of War

TACCSF	Theater Air Command and Control Simulation Facility
TBM	Theater Ballistic Missile
VV&A	Verification, Validation, and Accreditation
WAN	Wide Area Network

1. Introduction and Purpose

Recent technological advances have enabled geographically distributed sites to share a "synthetic battlefield" with a mix of live, virtual, and constructive simulations.[1] This sharing is called Advanced Distributed Simulation, referred to as ADS. While far from a mature technology, ADS offers great promise, and has already been embraced by the Science and Technology (S&T) community for demonstrating the potential of new technologies, and the training community for (among other things) training decisionmakers at virtually every level of the military decisionmaking hierarchy. This report considers the use of ADS to enhance the analysis process. Our hope is to assist those in the vanguard of using ADS for analysis.

Background and Motivation

The analysis community has been slow to embrace ADS,[2] partly because of the professional skepticism that is essential to the analytic process, the high cost of today's ADS, plain old inertia, and a wary reaction to the excessive enthusiasm of some of the technologists who have enabled ADS. There is far from a consensus yet on how (if) ADS can be used for analysis. Many analysts are mired in the belief that the only "scientific method" for analysis requires detailed computer models and/or statistical analysis of massive databases. However, there is a large segment of the analytic profession that recognizes the need for the analysis community to try to understand the potential contribution of ADS, and to try to exploit that potential.

While the initial efforts have emphasized technology demonstration and, to a lesser degree, training, the analytic community should take advantage of these pioneering efforts and learn from them. Accomplishing this end will require that analysts reorient to the evolving paradigm, and the attitude adjustment necessary to leverage existing ADS demonstrations. A major part of this process is the education/reeducation of the analysts. Equally important is the reeducation of the decisionmakers who commission ADS-based studies, and then subsequently turn the results into policy.

---

[1]See Kerchner, Friel, and Lucas (1996).

[2]As discussed by Dr. Anita Jones, Director, Defense Research and Engineering, in her keynote address at the 1996 MORS Workshop on Advanced Distributed Simulations for Analysis (ADSA).

Even though ADS is still in its infancy, ADS for analysis is rapidly becoming a reality and may affect important decisions. Pioneering analysis efforts are already under way, such as the Airborne Laser (ABL) tests, the Anti-Armor Advanced Technical Demonstration (A2ATD) experiments, and a joint Air Force and Navy effort to study whether and how the Cooperative Engagement Capability (CEC) should be extended to the Airborne Warning and Control System (AWACS). These efforts are discovering both a high potential benefit and a steep learning curve associated with ADS for analysis.

With this analyst's guide, we hope to acquaint both potential users and consumers with ADS so that it is used properly, and does not suffer from excessive hype followed by dashed expectations. To develop the analysis framework, we have synthesized lessons from early experiences and combined them with principles acquired from decades of traditional analysis. In fact, one of the findings in Lt. Col. Bob Sheldon's working group at the Military Operations Research Society (MORS) ADSA conference was that "ADS is not a new methodology for analysis—all the components, including human-in-the-loop (HITL), have been available to analysts for years. Rather, it is glue that allows analysts to tie together formerly stand-alone analytic tools into a more powerful mechanism for modeling complex problems." Therefore, in highlighting the special analytic needs of ADS it is sometimes necessary to repeat what experienced analysts are already familiar with.

The need for a guide such as this is well recognized by analysts, as evidenced by the extensive discussions at the 1996 MORS ADSA workshop and semi-annual Distributed Interactive Simulation (DIS) meetings. Two related efforts are the Analysis and Engineering (A&E) Tiger Team report,[3] led by Dr. Dale Pace, and the DIS workshop Exercise, Management, and Feedback (EMF) Rationale Document,[4] led by Mr. William Tucker. Our role, with respect to these other efforts, puts a greater emphasis on the analysis process and how ADS fits into it, rather than replacing it. This report contains some redundancy with the other ones; when possible, we will refer to them rather than replicating the information they contain. In particular, the A&E document will contain more detail on engineering-level models and the EMF paper has much more information on controlling and executing large distributed ADS exercises.

---

[3]See the Institute for Simulation Training, *Analysis & Engineering (A&E) DIS ++ Application Guidelines*, 1996.

[4]See the Institute for Simulation Training, *Exercise Management & Feedback (EMF)—Rationale*, 1996.

Intended Audience

There are three primary audiences for this research: decisionmakers; technologists, modelers, and trainers; and analysts.

Decisionmakers

Ultimately, the value of an analysis must be judged by those who use the analysis to inform a decision or communicate that decision to others (for example, if and how the CEC should be employed on AWACS). The newness of ADS is such that those who might sponsor or consume such a study may not fully appreciate all the accompanying consequences. A better understanding of what ADS can and cannot do, as well as its roles in analysis, should be useful to decisionmakers in determining whether and how an ADS component should be part of an analysis they commission. Furthermore, this understanding will help decisionmakers interpret and communicate ADS analysis results. Of particular interest to this audience are Section 2 (What Is ADS and Its Roles in Analysis) and Section 3 (High-Level Design).

Technologists, Modelers, and Trainers

The vast majority of the ADS community consists of technologists, modelers, and trainers. Part of the promise of ADS is that it will help with analysis, particularly with acquisition.[5] A better understanding of analytic requirements will help this broader ADS community direct their efforts to improve the analytic utility of ADS. Moreover, many of the organizations involved in the early analysis efforts have traditionally emphasized training and/or technology. The lessons and principles that analysts have learned over the years, detailed here, might help ensure the quality of their analysis contributions. Of particular interest to these groups will be the sections on designing ADS experiments (see Sections 3 and 4 and Appendix A), integration and testing from an analysis perspective (Section 6), and post-exercise analysis (Section 8).

Analysts

The analytic community is trying to understand if and how they should use ADS. This report synthesizes lessons from both traditional analysis and empirical

---

[5]*The DIS Vision* states that ADS will "transform the acquisition process from within," Institute for Simulation and Training, DIS Steering Committee (1994).

observations from initial ADS analysis efforts that may help analysis directors and their staffs in (1) determining the proper role of ADS in an analysis, (2) designing, testing, conducting, and analyzing ADS and traditional experiments, and (3) communicating the analytic conclusions. The greatest benefit from this report should accrue to those organizations that are new to ADS and can thus most benefit from others' hard-earned lessons learned. Some may be familiar with parts of this material, so we emphasize those aspects that are particularly important to the elements of ADS that are the most different from traditional practice—that is, large distributed HITL ADS.

Report Context

This report covers a broad range of topics at variable resolutions of detail. The following subsections discuss two "context areas" that may help readers better track and interpret this research.

Goals and Coverage

This guide's goal is to help readers better understand some critical analysis topics when considering the use of ADS for analysis or interpreting ADS analysis products.[6] Because it is impossible to comprehensively treat all of the subjects discussed, we emphasize the breadth of analysis issues over depth in their coverage—with references to more detailed resources for those desiring more information. Moreover, the depth of coverage is highly variable. The greatest detail is provided on the roles of ADS in the analysis process and experimental design.

ADS Environments Emphasized

ADS environments span the gambit from dynamically linking two simulations within one site to a multisite global simulation involving scores of models and humans-in-the-loop. Our default discussion emphasizes large distributed and HITL simulations, for several reasons. Primarily, those ADS simulations that are closer to the former[7] are the most like the current situation and thus most familiar to experienced analysts. Furthermore, the more modest constructive-only ADS simulations are a "lesser included" case of the larger more-distributed

---

[6]Furthermore, for those new to ADS for analysis, this guide is an introduction to analysis issues they might consider.

[7]That is, closer in terms of a small number of models, lack of HITL components, and few distributed sites (and thus there is a significant amount of local expertise and control).

efforts with HITL components. Additionally, most of the high-profile efforts involve multisite HITL simulations. Finally, many of the most important potential benefits and analysis challenges occur when linking multiple (sometimes joint) distributed simulations with HITL participants (see Section 2, What ADS Can and Cannot Do).

Outline

Section 2 defines ADS and discusses its roles in an analysis, including using ADS within a broader analysis context. This approach takes advantage of the benefits ADS provides and uses traditional constructive models to complement its weaknesses, focus the limited ADS experiments, and extend the results. Sections 3 and 4 discuss designing the ADS experiments—the area closest to traditional analysis. In these sections we emphasize points that are particularly important in large distributed HITL ADS exercises. Sections 5 through 8 contain material on exercise preparation, integration and testing, exercise management, and post-exercise activities vital to analysis. Sections 9 and 10 touch on two issues confronting ADS for analysis: verification, validation, and accreditation (VV&A) of distributed simulations and conducting analysis in training or operational exercises. Section 11 presents our general conclusions. Appendix A, which provides more details on designing experiments, is intended for simulation practitioners without an extensive analysis background. Appendix B contains our findings on activities the analysis community can take to improve ADS analysis capabilities.

2. What Are ADS and Its Roles in Analysis?

To understand how ADS contributes to military analysis, it is necessary to first understand the role of combat modeling and simulation (M&S). Defense analysts provide information to decisionmakers who must make policy decisions that consume billions of dollars and may affect many lives. The ongoing Quadrennial Defense Review (QDR), which will influence how we structure our forces (and thus our security) in an era of declining budgets, is one such example. Many defense scenarios, systems, and tradeoffs are too complex, with unpredictable interactions, for humans to reliably understand and/or reason through. Furthermore, there is little, if any, data that can be used to affect many of these decisions. For example, no one can fly a B-2 over Moscow to see if it will be able to penetrate the year 2010 Russian air defense system. Thus, simulation models—abstractions of reality—are used to gain insights and otherwise help inform these vital decisions.

ADS technologies, such as networks, architectures, formal standards, and protocols, allow multiple disparate simulations, simulators, systems, and people to share a virtual battlefield. ADS and other M&S can thus potentially be used to improve and explain decisions regarding[1]

- force structure,
- force modernization,
- mission planning,
- requirements definition,
- force employment,
- doctrine, and
- tactics.

---

[1]The analytic potential of ADS is not limited to this list.

Definitions of ADS, DIS, and High-Level Architecture (HLA)

There is no consensus definition of ADS, or even DIS, as witnessed by the debate at the MORS workshop on ADS for analysis. Here we present several community definitions on these and related topics. We take a very broad view of ADS: Our definition below (Kerchner, Friel, and Lucas) is based on the January 1993 report of the Defense Science Board Task Force on Simulation, Readiness, and Prototyping. This definition subsumes those of DIS and DIS++.

Definitions of ADS

Kerchner, Friel, and Lucas, in *Understanding the Air Force's Capability to Effectively Apply Advanced Distributed Simulation for Analysis*, RAND, MR-744-AF, 1996, define ADS as

> The ADVANCED enabling technologies [such as networks, architectures, formal standards and protocols] that allow geographically DISTRIBUTED sites to share a "synthetic battlefield" with a mix of live, virtual, and constructive SIMULATIONS.

In this report, the relationship of ADS to DIS is defined as:

> Another, older, term for these technologies and vision is Distributed Interactive Simulation (DIS). However, DIS is now used to refer specifically to ADS in the context of a specific set of standards and protocols, including IEEE 1278. Thus, ADS is broader than DIS, for it includes distributed simulations such as the Aggregate Level Simulation Protocol (ALSP) confederations, that do not conform to DIS standards.

Our definition follows from how we see ADS being used in practice. Throughout this report ADS should be interpreted as such.

The MORS workshop on Advanced Distributed Simulations for Analysis (ADSA '96), defines ADS as

> The evolving DoD Distributed Modeling and Simulation infrastructure, including synthetic environments, run-time infrastructures, and connected human-in-the loop simulations such as DIS.

Sikora and Coose, in "What in the World is ADS?" *PHALANX*, Vol. 28, No. 2, June 1995, write:

> The technology area that provides a time-coherent, interactive synthetic environment through geographically distributed and potentially dissimilar simulations is called, reasonably enough, Advanced Distributed Simulation (ADS). The distributed simulations can be any combination of real people,

real equipment, or computer programs which simulate people, equipment, and their interactions.

Definitions of DIS

From *The DIS Vision: A Map to the Future of Distributed Simulation*, Institute for Simulation and Training (IST), May 1994, version 1:

> a synthetic environment within which humans may interact through simulation(s) and/or simulators at multiple networked sites using compliant architecture, modeling, protocols, standards, and data bases.

In *Quality Research*, 1995, the second part of the definition is the same as above except "and/or simulators" is deleted and "networked sites" is "sites networked."

> (1) Program to electronically link organizations operating in the four domains: advanced concepts and requirements; military operations; research, development, and acquisition; and training. (2) A synthetic environment within which humans may interact through simulation(s) at multiple sites networked using compliant architecture, modeling, protocols, standards, and data bases.

From Department of the Army, *DIS Master Plan*:

> The essence of DIS is the creation of a synthetic environment within which humans and simulations interact at multiple networked sites using compliant architecture, modeling, protocols, standards, and databases.

The High-Level Architecture

The Department of Defense is working hard on facilitating M&S interoperability and reuse, as demonstrated by the following quote.[2]

> In accordance with the DoD Modeling and Simulation Master Plan (DoD 5000.59-P, 1995), the Defense Modeling and Simulation Office (DMSO) is leading a DoD-wide effort to establish a common technical framework to *facilitate the interoperability* of all types of models and simulations among themselves and with C4I systems, as well as to *facilitate the reuse* of M&S components. This common technical framework includes the High Level Architecture, which represents *the highest priority effort within the DoD modeling and simulation community*.

In fact, the HLA has been designated as the standard technical architecture for *all* DoD simulations.[3]

---

[2]This quote was gleaned from the DMSO web site, http://www.dmso.mil/projects/hla/

[3]It was made official in a letter from Paul Kaminski, Under Secretary of Defense for Acquisition and Technology (September 10, 1996) to senior military leaders.

From the DoD, *The High Level Architecture rules, version 1.0*:

> The High Level Architecture (HLA) is defined by a set of rules, an interface specification, and an object model template (OMT). ... The overall objective of the DoD common technical framework, which includes the HLA, is to support interoperability and reuse. The HLA provides the structural basis for interoperability; most of the rules ... have been included for that reason.

Key Distinguishing ADS Characteristics Versus Traditional Analysis

Models inevitably employ simplified representations of the actual environment, and their results must be interpreted keeping the implications of the simplifications in mind. ADS can in some cases remove or reduce simplifications that are present in constructive, non-distributed M&S. For example, virtual or live human participants in an ADS environment remove simplifications associated with the constructive representation of human participants. Distributed access to authoritative models of certain systems can reduce the simplifications otherwise associated with representing those systems, as well as save time and avoid errors. For example, the Air Force, Army, and Navy can all represent their own systems.

Analysis in an ADS environment can be very different from what we shall call *traditional* analysis, that is, analysis using a stand-alone constructive model, or stand-alone constructive models run in sequence. This section highlights those differences by describing two likely ADS analysis environments.

ADS Analysis with a Constructive Model Federation

Analytic Example: The JANUS Federation. It is very common in defense analysis to have difficult questions to answer in a short period of time. This difficulty is compounded when the analysis team believes their primary analysis (simulation) tool is deficient in functional capabilities that cannot be incorporated within time and budget constraints. A few years ago, this situation confronted RAND researchers involved in a series of studies on combined arms land combat, such as modernization and employment strategies relating to rapid force projection. The analysis team employed a variety of models, with JANUS the most used for this type of study.

Driven by analytic requirements, before it was fashionable, the RAND analysts decided to create a model federation built around the JANUS model. This was largely motivated by the fact that many portions of JANUS, such as the modeling of aviation assets and the detection of low-radar-cross-section targets, did not adequately represent new phenomena, but other simulations did.

The first step in the process was to create the Seamless Model Interface called SEMINT. Unlike the DIS philosophy of generalized protocols, SEMINT was designed to communicate only among the specific models that had been selected for the federation. DIS protocols had not yet been defined. SEMINT acts as a run-time library for all of the models' objects (entities) and updates each of the connected models on the status of each object at specified intervals. The requests for other model inputs are originated in JANUS and transmitted via SEMINT; JANUS is in a state of suspension until it receives a response from the other models via SEMINT. SEMINT also ensures that each of the models stays in synchronization.

The models in the JANUS federation include the Cartographic Analysis and Geographic Information System (CAGIS), which provides representation of the terrain upon which all the other simulations move their elements. CAGIS uses digital terrain data from the Defense Mapping Agency (DMA), as well as features data from DMA and from surveillance satellites. In addition, the RAND Jamming and Radar Simulation (RJARS) is included in the federation to provide attrition calculations for rotary and fixed-wing aircraft operating over the terrain provided by CAGIS. The RAND Target Acquisition Model (RTAM) calculates the acquisition of low-observables targets on or above the CAGIS terrain. This computation uses dynamic background and signature data from CAGIS as a function of terrain and location.

Another member of the federation is the Model to Assess Damage to Armor by Munitions (MADAM). This model evaluates the damage caused by smart and brilliant cluster munitions against armored fighting vehicles and trucks operating on CAGIS terrain. MADAM includes an acoustical model that locates artillery for counter-battery fire and armor for non-line-of-sight engagement. The acoustical model is also involved with the use of smart mine fields. When JANUS requires a determination of an input, the requirement is processed through SEMINT to the model that will perform the calculation. The results are then passed back through SEMINT to JANUS.

By allowing JANUS to use information from more detailed models, the analysis team feels it can credibly address a host of important analysis issues it previously could not. We emphasize this example for it is the most mature analysis federation we are aware of. Indeed, there have been many thousands of production runs. The JANUS federation has been used as the primary modeling tool in many studies for RAND clients, including the Rapid Force Projection

Technologies Project conducted in support of the Rapid Force Projection Initiative, the Future Close Support Study for the Office of the Secretary of Defense (OSD) Defense Program Analysis and Evaluation (DPA&E), and the Military Applications of Robotic Systems for the Defense Advanced Research Projects Agency (DARPA).[4]

There are attributes that contribute to the success of the JANUS federation that may not always apply to general ADS federations. First, the federation was assembled by analysts at a single location and consisted of trusted and well-known models. Second, the models are explicitly integrated by an active controlling agent (SEMINT) rather than just exchanging state information, as is done in DIS federations. Finally, the analysis team had extensive experience with all of the components of the federation and were able to put it through testing and operational use over a lengthy period of time.

General Discussion. Stand-alone constructive simulations are a mainstay of current analysis. Their use is well understood by the analysis community, and analysis with a stand-alone simulation offers advantages of local control, although pressures for model standardization in practice often limit the ability of the individual analyst to freely modify a model or its data. In contrast, ADS implies the use of cooperating federations of simulations.

Federations whose components consist entirely of constructive simulations (federates) are comparable to stand-alone constructive simulations with regard to the analytic strategies they support. Constructive federations are built from constructive federates that together provide a representation of the relevant parts of the combat environment. ADS federates will normally run in parallel, with the potential for capturing interactions between any pair of federated simulations (in any direction). We specifically exclude federations that are run sequentially. Sequential federations use multiple simulations, but the results of one are used by the next without the possibility of dynamic interaction. For example, one simulation might compute aircraft flight paths that are used by a second simulation to compute detection and tracking probabilities, but interactions arising from detection/tracking that might cause the aircraft to alter their flight paths are not treated. In the dynamic ADS environment, this interaction is captured by passing state information between the two simulations.

When the federates are physically located locally, this environment will be most similar to that of the traditional analysis. However, there will still be major

---

[4] Some examples of the many analysis products using the JANUS federation are Matsumura et al. (1997), Don et al. (1997), and Steeb et al. (1996).

12

differences between a local ADS environment and the constructive federations occasionally seen today. The most striking difference will result from the interface standardization implicit in ADS. This standardization should permit analysts to assemble the assets needed to simulate a scenario of interest, at an appropriate level of detail, with far less cost and effort than that associated with either (1) augmenting a constructive model to give it the needed capabilities or (2) building a federation without the benefit of preexisting standards. However, as we will discuss later, this does not imply that ADS will permit meaningful analysis to be performed simply by plugging simulation components together.

ADS' standardization of interfaces extends to post-processors, including visualization tools. Although analysts often neglect the presentation of findings, their job includes this facet, and it is critical to the objective of having an analysis make a difference in the decisionmaking process. The easy ability to plug into preexisting and highly capable visualization and analysis tools will enormously enhance this aspect of the analysis process.

Many ADS constructive federations will include geographically distributed federates. This will be advantageous when local computer resources are inadequate to support all the federates, or when individual federates require expert support that is not available at the analyst's local site. One of the most important advantages in this area will be in the arenas of (1) analysis that is a joint service effort and (2) analysis that supports decisions with joint implications. The acceptability of these efforts will require the use of system representations acceptable to the services that own each system, and the ability of a geographically distributed federation to incorporate such acceptable simulations, wherever they reside, is a key to success of such a federation.

ADS Analysis Including Human-in-the-Loop Federates

Analytic Example: Should AWACS Get the CEC? Consider the following (partially fictional[5]) potential decision that could benefit from analysis. Senior military leadership wants to know what (and if and how) role an AWACS should play in an integrated joint air defense architecture. A major concern is coordinating the command and control roles between Air Force, Navy, and Army elements, with an emphasis on the ability of humans in the AWACS to perform both their Air Force and (new) joint missions.

---

[5]This example is inspired by the Joint Engagement Technology Study (JETS). The TACCSF and NRaD were the primary simulation sites (see Glossary).

A team charged with conducting an analysis informing this decision finds that there is no existing simulation that has the balance and detail it requires across the services' systems. This includes the very real concern of the services that their systems be represented "properly."[6] Furthermore, existing constructive simulations are known not to represent human perceptions, workloads, and decisions very well. These are all essential to the study. Moreover, while the importance of the decision has resulted in generous financial resources, the short time allowed makes it infeasible to augment an existing model (or facility)—or build a new one.

In such a case, ADS technologies offer hope—though not without challenges. In theory, the analysis team can assemble a distributed simulation[7] with Army, Air Force, and Navy simulations representing their own systems and active duty personnel in simulators representing some of the critical human command and control elements.

General Discussion. A generally acknowledged weakness of constructive simulations is their limited ability to provide realistic representations of the human decision processes that are an important part of combat. Analysis with constructive models is made more difficult when it becomes necessary to modify the analysis strategy to compensate for the limitations of the simulations, or when it is difficult to validate those human behaviors that are included.

These weaknesses can be mitigated by the use of human-in-the-loop federates, in the form of humans in simulators (virtual federates), or live entities who are instrumented to permit them to interact in an ADS environment. The inclusion of human players will generally provide more realistic decisions than when only constructive components are present. There is also the pragmatic advantage of the superior believability associated with HITL. Whether justified or not in a particular case, the ability of HITL to "sell" an analysis should not be underestimated.

However, we do not mean to imply that one can base an analysis solely on the use of ADS experiments involving HITL *instead of* experiments involving constructive runs. ADS experiments involving HITL are rarely adequate, by themselves, to meet the analysis needs. The most compelling (but by no means the only) reason is that the number of replications possible with HITL is, with rare exceptions, going to be limited to a few tens. On the other hand, analysis

---

[6]Of course, one must always be wary of unintended biases that favor one service's systems relative to those of another.

[7]Assuming there exists a library of appropriate compatible simulations, such as those that are DIS compliant, for example.

requirements to explore a wide variety of conditions and to obtain good statistical precision can dictate the need for hundreds, and often many thousands, of runs. This can be achieved only through the use of constructive runs. Such runs are needed, then, but if they are to have maximum credibility, it is important that the constructive M&S representations of critical human performance factors are compared with, and calibrated against, ADS runs involving HITL. ADS experiments can be designed to use virtual (or live) simulation to inform analysts regarding human factors, for instance, to identify critical factors and ensure that they are in fact accounted for in the constructive models.

This point leads naturally into a second way in which ADS experiments involving HITL can inform analysis. All experienced analysts realize that constructive models rarely incorporate all of the relevant qualitative factors that influence combat, despite ample evidence of their importance. This point pertains to any representations, even calibrated or validated ones, and the omission is likely to remain even when the relevant qualitative factors are explicitly identified. While much progress has been made in the area of modeling qualitative factors (Davis, 1989; Dupuy, 1987), their incorporation into constructive M&S remains extremely difficult, and is unlikely to be adequate in the near term. One way to improve an analysis based on constructive models is to use ADS runs with HITL participants to incorporate qualitative factors that would otherwise be missing or poorly treated.

We have just pointed out that for many analysis problems, relying exclusively on HITL runs is unrealistic, so the idea is to make the best use of the limited number of HITL runs available. Use constructive runs to identify key or critical cases, and explore these with ADS experiments utilizing HITL. An overall approach in which ADS HITL runs and constructive-only (possibly ADS) simulation mutually support each other is a new way of doing business that will be outside the direct experience of most analysts.

What ADS Can and Cannot Do

When considering whether ADS is an appropriate component of an analysis, the study director must think about the advantages and challenges associated with using ADS. Here we review some key advantages and challenges for both constructive-only federations and those involving humans-in-the-loop.

Generally speaking, ADS facilitates the combining of multiple disparate simulations into a single joint simulation, thereby allowing us to simulate joint theaterwide scenarios at previously unobtainable levels of detail. Furthermore,

ADS makes possible more realistic treatment of human performance, a notable weakness with constructive computer simulations. Additionally, multidiscipline research teams that explicitly include warfighters will have enhanced credibility. However, there is no free lunch. The additional costs associated with using ADS (over traditional analysis) generally relate to the difficulty associated with obtaining the degree of reliability and control one can obtain in a well-controlled laboratory experiment.

It is important to realize that the advantages are not automatically realized and that the challenges can be mitigated—technology trends should help in both areas. Moreover, while these advantages and challenges are discussed in the context of analysis with ADS, they are not necessarily unique to it.

Potential ADS Analysis Advantages

In theory, the potential benefits of ADS-supported analysis follow from the vision for DIS as stated in *The DIS Vision* (1994):

> The primary mission of DIS is to define an infrastructure for linking simulations of various types at multiple locations to create realistic, complex, virtual "worlds" for the simulation of highly interactive activities.

To the extent this capability is achieved, it can dramatically improve on the weaknesses of traditional analysis.

Advantages Achievable in Constructive-Only Federations

- **Provide an ability to simulate larger and higher-resolution scenarios.** By facilitating massively parallel simulation, ADS allows analysts to include important factors in large scenarios, such as environmental effects, that are not feasible—even with supercomputers—in stand-alone constructive models. As with any constructive model, these can be run at faster or slower than real time.

- **Facilitate the use of "more authoritative" models and expertise, i.e., those developed by the experts.** ADS allows sites to tap into the tools and expertise of geographically distant sites. The acceptability of simulation-supported analysis efforts often depends on the credibility of the underlying model representations. This includes the data as well as the algorithms and processes that are contained in the simulations.

- **Facilitate joint analysis.** The acceptability of model-supported analysis efforts often requires the use of system representations acceptable to the military services that own each system, and the ability of a geographically distributed federation to incorporate such acceptable simulations, wherever they reside, will be a key to success in joint studies.

- **Provide a superior ability to communicate results.** Many analysis consumers better absorb and believe the analysis results when they are presented with visual displays and HITL simulators. While not unique to ADS, the "visual" *emphasis* in ADS can be critical to the objective of having an analysis make a difference in the decisionmaking process.

- **Achieve faster model development.** ADS standards make it theoretically feasible, though still challenging, for analysts to assemble the assets needed to simulate a scenario of interest, at an appropriate level of detail, with far less cost and effort than it takes to augment a constructive model to give it the needed capabilities.

- **There may also be security advantages.** For example, a particular federate may include highly classified representations of national surveillance assets, but the outputs of that federate take the form of simulated communications that are sanitized to a lower security level. The physical location of the federate at a high-security site may be the only viable way of including the federate in the federation.

Additional Advantages That Can Be Accrued with Humans and Hardware in the Loop

- **More realistic treatment of human interactions and performance limitations.** Traditional simulations are "woefully inadequate in this area."[8] Human interactions are especially vital in simulating information aspects of combat, such as C4ISR.[9] This may also be particularly valuable in simulating highly reactive three-dimensional combat situations, such as urban warfare.

- **More realistic treatment of hardware.** By facilitating the inclusion of hardware-in-the-loop, ADS allows one to simulate real system performance and reliability.

- **Facilitates warfighter input.** Warfighters can be in-the-loop early and continuously throughout the study. This enhances process visibility and allows for thorough warfighter feedback—helping accrue "face validity." Having warfighters imbedded within simulations is valuable in obtaining insights into the cause-and-effect "drivers" of the simulated combat and can

---

[8]A. Jones, "ADS for Analysis," *PHALANX*, Vol. 29, No. 2, June 1996.
[9]See C. Marshall and R. Garrett, "Simulation for C4ISR," *PHALANX*, Vol. 29, No. 1, March 1996.

be extremely useful when developing tactical concepts to enhance or defend against new weapons systems.

Primary ADS Analysis Challenges

ADS is a new analytic paradigm. As with any new paradigm, there are challenges to be overcome. It requires a new way of doing business. As Dr. Stuart Starr reported at the June 1996 MORS conference, "the challenges associated with ADS transcend the abilities and resources of the individual analyst."

Challenges Posed in Constructive-Only Federations

- **Understanding limitations and biases.** The synthesized joint synthetic battlespace is immensely complex. The combining of multiple disparate simulations and databases, built by different people for different purposes, greatly exacerbates the difficulties associated with understanding the limitations that are inherent in any simulation or combination of models.

- **Plug and play is currently an illusion.** Experience has shown us that compliance to standards does not ensure interoperability. "Even defining what interoperability for a large, distributed simulation means has proven to be a Herculean task."[10]

- **Exercise reliability suffers.** This is a consequence of having more elements together with fewer opportunities to "iron out the bugs." Reliability is especially important for situations where system crashes can be catastrophic to analysis measures—for example, a crash that invalidates the run's loss exchange ratio.

- **Network latencies[11] can cause difficulties.** Network latencies make exact reproducibility, vital for error checking and cause-and-effect tracing, impossible. Furthermore, extreme latencies may affect the validity of some types of interactions, such as air-to-air combat.

- **Verification, Validation, and Accreditation are particularly difficult.** The history of VV&A of stand-alone constructive combat simulations contains few success stories. Any federation, particularly a distributed one, will exist in the desired state for only a short time. There will be insufficient time for rigorous validation, much less the nine-step validation process as outlined in

---

[10]R. Garrett, "Architectural Design Considerations," *PHALANX,* Vol. 29, No. 2, June 1996.
[11]Delays in data exchanged between models on the network.

the VV&A section of the *Exercise Management and Feedback* recommended practices guide (Institute for Simulation and Training, 1995).

- **Current manpower and dollar investment requirements are high.** It is said that if someone has undertaken an ADS-supported study he must have a big budget. The inclusion of multiple sites and the associated additional tasks that must be done simply requires more work than traditional analysis. Furthermore, if it is HITL, large personnel resources are necessary.

- **Restrictions associated with the use of proprietary models.** Proper use may require the use or sharing of simulations and data that are considered corporate assets or service eyes only.

- **Data output is large and distributed.** This can be mitigated by carefully designing the data to gather—rather than trying to get everything.

- **Standards and protocols may not produce the type of data that are needed.** Current protocols, such as DIS 2.X, contain "what" information, such as position coordinates, and not much "why" information, such as situational awareness. Often, analysis will require that provisions be made to acquire additional information at local sites.

- **There may be security challenges.** The security classification level may have to be restricted to that of the lowest network or site limit in the federation. Care must be taken to ensure that no person can infer from data on the network or a site classified information higher than that assigned.

Additional Challenges with HITL

- **Inability to take large samples.** Real-time constraints associated with HITL exercises restrict the number of samples one is able to obtain, which limits the number of inputs that can be varied. It is thus difficult to vary many research variables and to screen for possible confounding factors. It is also difficult to get the sample sizes necessary for precise statistical comparisons and estimations.

- **Human experimental difficulties.** There are a host of problems associated with experiments using humans. These include ensuring that the study sample is an adequate representation of the population about which inferences are to be made, participant learning and gaming, participant boredom, and exact reproducibility.

- **Visual displays are often immature or unbalanced.** Current real-time visual displays, although improving dramatically, are quite immature. Furthermore, visual displays vary from site to site. Such imbalances may bias simulation outcomes.

What the Advantages and Challenges Suggest

The above potentials and weaknesses with (virtual) ADS suggest the following set of ideas.[12] Treat large HITL simulations as a few high-value experiments, and traditional constructive simulations as tools to focus, extend, and complement the few HITL experiments.[13] At a high level, use HITL ADS experiments to

- evaluate issues or hypotheses that can be effectively analyzed in virtual tests,
- calibrate human performance factors in constructive models, and
- provide insights and cross-check constructive model results.

Use traditional constructive models to

- focus (design) limited ADS runs on the most important cases, and
- perform the bulk of the exploration (after calibration by ADS).

ADS Within a Broader Research Plan

ADS Analysis Including Human-in-the-Loop Federates, above, outlined an approach to using ADS in analysis that includes using ADS experiments with HITL to calibrate constructive models, and constructive models to identify key cases for examination with ADS/HITL. We discuss this approach further here. It is illustrated in Figure 2.1.

In the center is the activity of designing, building, and *calibrating* constructive M&S. This activity, which is ongoing and highly iterative over a period of many years, often draws upon either small specialized ADS experiments (top right), as discussed in the last subsection, or even larger distributed war games (top left) for both insights and data. Further, this activity may inform the design of those experiments in the first place. The connection between the ovals "Conduct small, specialized ADS HITL exercises ..." and "Analyze problems" refers to the direct contribution of ADS runs to analysis. As noted above, such analysis would normally be in conjunction with additional runs using constructive M&S. And,

---

[12]Callero, Veit, Gritton, and Steeb (1994) have an excellent related discussion about integrating simulators with other tools to enhance weapon system analysis in the acquisition process. They have a strong emphasis on validity issues.

[13]The same idea also applies to situations where constructive-only ADS federations can be focused, extended, or complemented by stand-alone constructive simulations. This will be especially valuable when the ADS federation is constrained in the number of cases that can be run.

20

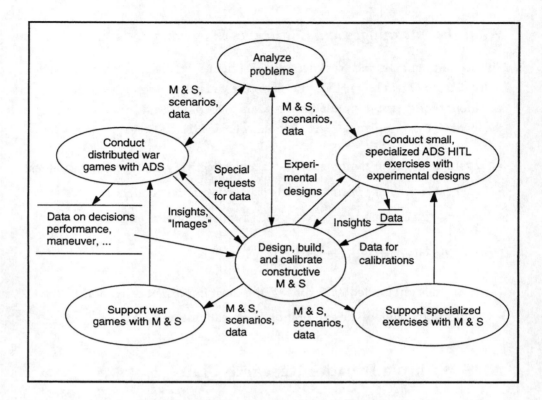

Figure 2.1—ADS-Mediated Interactions Among Model Building, Analysis, Experimentation, and Training

to make things more complicated but realistic, the M&S will typically be supporting and included in the experiments and exercises. This is not a simple linear flow, but it is perhaps the image of how we should view the continuing organic processes of studying, innovating, experimenting, training, planning, and analyzing if they are increasingly interwoven.

Figure 2.2 shows how this concept looks in the context of performing an analysis in which ADS plays a role.[14] It is idealized in a number of respects, including the assumption of adequate time and resources to include all the steps iteratively, but the *interplay*[15] shown can be usefully looked at as an ideal to approach as much as is practical.

---

[14] In outlining this analysis, we emphasize the role ADS plays in providing superior treatment of human performance factors. We did so because this is one of the most important roles ADS can play. It is impossible to list all such roles.

[15] We highlight the interplay between the different types of simulations. We feel such an interplay would benefit many of the early ADS analysis efforts we have observed. This is not meant to imply that there are no analysis studies that can credibly use either HITL ADS simulations or stand-alone constructive models. For example, for some human factors studies there may not be much benefit from stand-alone constructive runs.

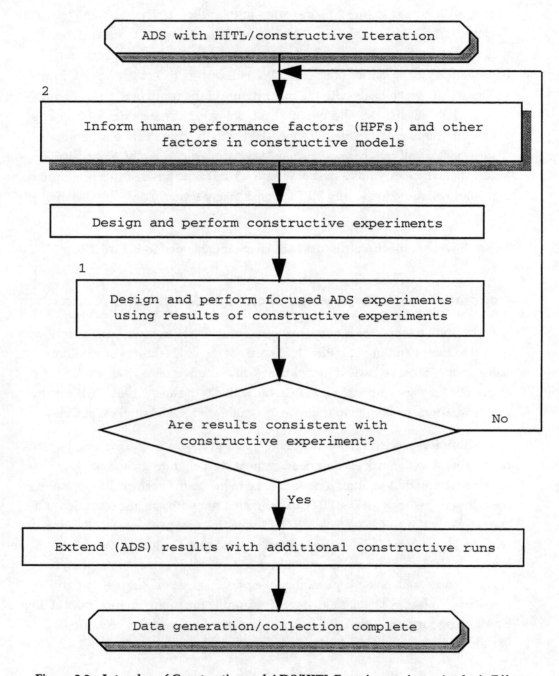

Figure 2.2—Interplay of Constructive and ADS/HITL Experiments in an Analysis Effort

In fact, ADS is used in three distinct ways in this scheme. The "Design and Perform" ADS experiment block (block 1) in Figure 2.2 refers to the use most visible to the consumers of the analysis, and corresponds to the collection of baseline runs for scenarios of interest. We will refer to this as *baseline ADS runs*. In Figure 2.3, which deals with the calibration[16] of the constructive simulations used in the analysis, a second use of ADS is shown: HITL experiments aimed at calibrating human performance factors (HPFs) in the constructive simulation (block 2). We will call these *calibration ADS experiments*. Finally, embedded in Figure 2.4, which expands a part of Figure 2.3, ADS can be used in a preliminary *exploratory* manner to identify HPFs that are likely scenario drivers (also in block 2).

The following subsections discuss each element contained in Figure 2.2.

Calibrate Human Performance Factors in Constructive Models

A first step in the process is to evaluate the constructive models, particularly with regard to their handling of HPFs. HPFs are not the only features of a military scenario that must be treated in the constructive models used in an analytic effort, but they are emphasized here because this is the area where ADS has the most (near-term) potential to improve our calibration/validation capabilities.

The calibration process is shown in more detail in Figure 2.3. We avoid the term *validate* because the process described emphasizes calibrating parameters in constructive models of human performance to match the performance of human operators in particular ADS/HITL experiments run with specific scenarios. Of course, calibrating a model to ADS/HITL data does *not* guarantee calibration to real-world performance.[17] Issues critical to validation, including the validity of the ADS/HITL experiment, are not addressed. Another critical validation issue, also not addressed, is the applicability of a particular set of parameters to other scenarios. In fact, we deliberately avoid assuming broader scenario applicability by allowing for a repetition of the calibration process when new scenario variations are used.

---

[16]Here calibration can refer to either data elements or model outcomes, whichever is appropriate. Differences between physical models make general model calibration very difficult, whereas taking data from HITL to calibrate/inform constructive models is relatively straightforward.

[17]In fact, for many tasks, such as air-to-air combat, the HITL environments may contain substantial biases, in this case perhaps because visual displays are very different from the real world (and unable to support and thus analyze typical tactics). In other cases, such as with AWACS command and control displays, the HITL environment may be close to the real world.

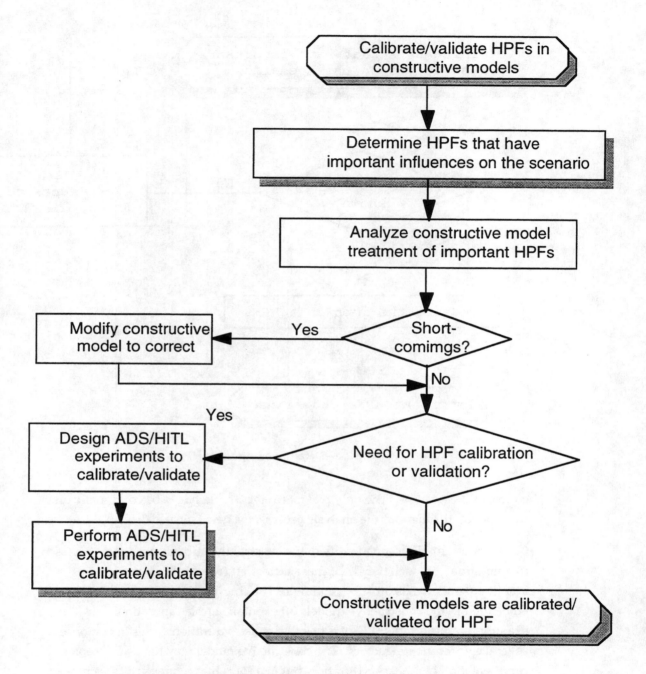

Figure 2.3—Calibrate/Validate HPFs in Constructive Models

A first step is to make sure that we understand what the important HPFs are. An expansion of this step, shown in Figure 2.4, reveals that it may be desirable to make an ADS experiment here to gain insight into what HPFs are scenario drivers. There is a subtle problem associated with such an experiment. The important drivers in the scenarios selected for the experiment may not be the important drivers in all the scenario variations that will be examined in the

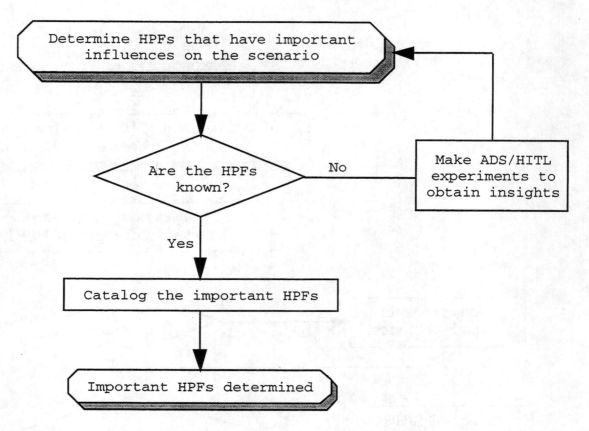

Figure 2.4—Determine Important HPFs

course of the analysis. The only real fix for the problem is to be continually on the lookout, throughout the analysis process, for new scenario drivers.

Once important HPFs are identified, the constructive models need to be assessed to determine how well they treat these factors. If (when) shortcomings are identified, the models must be modified. The changes to HPF submodels embedded in the constructive models often entail simple calibration. For example, an important HPF might be that a task's duration varies in response to several environmental factors. The constructive model may have an acceptable representation of the task's duration, but in a form that requires the specification of values of several numeric parameters. It is likely that the required data will not be available from preexisting sources. The data (along with other similar data needed by the constructive model) can be obtained by performing an ADS experiment with HITL to measure the values of the needed parameters (or values from which the needed parameters can be computed). We regard such calibration experiments as one of the most important analytic uses of ADS.

Note that it may not be possible to obtain calibration data by relying on databases gathered from, say, training exercises, even if they are ADS training exercises. Virtual and live components need to be instrumented to capture calibration data for constructive models, and in most cases prior knowledge of what data are needed is necessary to configure the instrumentation.

Design and Perform Constructive Experiments

Fundamental constraints such as limitations to real time, and likely pragmatic constraints associated with expense and schedule, severely limit the number of ADS/HITL runs that can be performed in the course of an analytic effort. It is thus essential that the ADS/HITL runs be viewed as a scarce and valuable resource to be used as efficiently as possible. This perspective implies the need for careful planning of the ADS/HITL runs.

A perspective we take, that is applicable to many analytic efforts, is that the analysis includes looking at scenario variants that can be thought of as points in a multidimensional parametric space. It is almost certain that the vast majority of this space is uninteresting in the sense it has no impact on the decision process that the analysis is supporting. For instance, if the issue has to do with survival tactics for a flight of fighter aircraft in a screening role, then uninteresting scenario variations include those where the enemy force is so overmatched that different tactics have no essential difference in outcomes.

Since ADS/HITL is a scarce commodity, it is clearly important to focus the ADS/HITL runs on the interesting scenario variants. Unfortunately, in complex situations it will not be obvious, *a priori*, which scenario variants are in fact interesting. Constructive simulation runs, aimed at assisting the ADS/HITL experimental design process by searching for interesting regions of the parameter space, can be invaluable in this situation.

Design and Perform ADS Experiments Using Results of Constructive Experiments

If ADS/HITL were not available, the analysis might proceed by making just such a set of exploratory runs, identifying the interesting scenario variations and then making a large number of runs in the interesting regions to obtain statistical accuracy and to understand how measures of interest vary as parameters are altered. In actual practice, this kind of preliminary exploratory modeling is rarely made. One reason is that constructive model runs are comparatively cheap; it is often easier to design an experiment assuming reasonable values for various scenario parameters, and then to reconsider and make additional runs if the results indicate they are needed. Brooks, Bankes, and Bennett (1997), and others, argue that this approach can miss important regions. However, their approach is beyond the scope of this guide, with its focus on ADS.

Another flaw in the above approach is that the interesting scenarios are likely to differ significantly from those used to calibrate the HPFs in the preliminary ADS/HITL experiments described above. Thus, the performance within the interesting regions is likely to be in error, as measured using the constructive models. This assumes, of course, that HPFs are critical drivers in the scenarios of interest.

This flaw can be addressed and corrected with another use of ADS/HITL. ADS/HITL runs, in the regions of interest identified by constructive models, can be used in two important ways:

1. They can confirm that the constructive model and the ADS/HITL runs are in substantial agreement. This will not be the case if the HPF models, as calibrated with one set of scenario variations, are not properly calibrated for use with the scenario variations of interest.

2. Insight gained from looking into the causes of the differences between the constructive runs and the ADS/HITL can identify HPFs that were not previously identified, but which play a critical role in the new scenario variations.

Either of the above insights can lead to additional calibration of the constructive models, and possibly to additional modification of the constructive models' treatment of HPFs. After this step is repeated, it may be necessary to repeat the exploratory constructive modeling phase to determine where the interesting scenario variations are located. It is expected that the outcomes of the previous set of ADS/HITL can be examined to decide if this step is in fact needed.

An additional point should be made. Our arguments have implicitly assumed that the differences between the constructive and ADS/HITL runs result from HPFs. This emphasis is appropriate because we want to focus on the ability of ADS to provide a superior treatment of important HPFs. However, the hardware models—the models of the physical performance of the military systems and of the environment[18]—may also differ between the constructive and the ADS simulations. In this case, differences between the hardware models must be resolved, although one cannot make any assumption about the superiority of either the hardware models used by the stand-alone constructive simulation or

---

[18]Such as weapon systems, sensors, command and control processing, terrain, etc.

the (constructive) hardware models used in the ADS/HITL runs. In contrast, it is implicit in the previous discussion (although not always correct) that the ADS/HITL simulation provides a superior treatment of HPFs than does the constructive simulation.

It can be a good idea to use the same hardware models in the constructive and ADS/HITL runs. This avoids additional convoluting factors, albeit at the risk of missing flaws in the hardware models that would surface in the form of discrepancies in the performance of the two versions of the model of a hardware system. If the "common hardware model" approach is used, the constructive model will generally not be a stand-alone model, but will in fact be an ADS constructive federation that is identical to the ADS/HITL federation except for the substitution of computer-generated forces (CGFs) for the HITL entities.

Extend ADS Results with Additional Constructive Runs

At a certain point, the consistency between the ADS/HITL and the constructive simulations will be acceptable for the purposes of the analysis,[19] and the ADS/HITL simulation will have provided a set of runs in the scenario regions of interest. This set will generally be limited in both statistical accuracy and in completeness of scenario excursions. The constructive simulation should now be used to extend the ADS/HITL results, both to get additional replications for statistical accuracy and to make excursions that determine the detailed relationship of output measures to variations in the scenarios.

This will nominally complete the data-gathering phase of the analytic effort, although we recognize that new questions are likely to arise as the analysis proceeds, which is likely to generate never-ending requirements for additional runs. This must be controlled by rational procedures, perhaps necessitated by time and money constraints.

---

[19]"Consistent for the purposes of the analysis" is a loaded phrase. If the ADS models' algorithms are different from the stand-alone ones, it may be essentially impossible to get exact agreement over a broad range of scenarios. How close is close enough must be evaluated on a case-by-case basis depending on how resulting conclusions are affected. For a discussion on integrating several different simulation types in an analysis, see Lucas, Bankes, and Vye (forthcoming).

3. High-Level Design

A high-level design is essential for any analysis effort. There are books that
address this topic much more comprehensively than we can, (e.g., see Quade,
1989). However, there are features of ADS that require special consideration.
This section touches on a few of the important high-level design issues that
deserve emphasis. The number of diverse participants in many ADS efforts and
the small number of trials magnify the importance of high-level design in large
distributed HITL analysis studies. Effective utilization of this process requires
participation from analysts, modelers, consumers, network managers,
developers, testers, and operationally experienced individuals. While the
breadth of ADS is such that there is no unique process, the following is
illustrative of the steps that are needed.

Formulation of Analytic Plan (Including Non-ADS Tools)

Problem Definition

The initial steps involve problem definition, including identifying the intended
beneficiaries of the analysis, the decisions to be made or issues examined, and the
insights to be gained.

Statement of Objectives, Goals, Critical Issues, and Questions to Be Resolved

In traditional analysis, a clear statement of the sponsor's objectives, goals, critical
issues, and questions to be resolved is essential. With the distributed nature of
ADS and the added burden of scheduling disparate organizations, facilities, and
Internet access, the need to thoroughly understand the goals and objectives of the
sponsor is essential. Otherwise, different organizations may have different
agendas.

What Information Is Required

What information is required from the experiments to make specified decisions,
resolve specific hypotheses, answer critical questions, and otherwise inform an

analysis? The information may come from ADS and/or other modeling sources, or previously run similar analytic efforts. While this information is required for any simulation-based analysis, many ADS efforts rely on only a few exercises with few opportunities after-the-fact to correct for inefficiencies with additional runs.

As with most analyses, top-level Measures of Merit (MOMs) such as FLOT (forward line of own troops) movement, loss exchange ratios, and survivability will be defined. Required Measures of Effectiveness (MOEs) and Measures of Performance (MOPs) will be derived from these in the detailed design. Broad scenario(s) requirements are critical and need to be specified here.[1] The high-level concepts of operations and rules of engagement must be defined and agreed upon by the distributed parties planning the exercise. This is particularly true when there is more than one uniform service involved.

Determine Analysis Strategy

A decision for any simulation-supported analysis effort is what strategy will be used to *reason* with simulation outcomes. Most ADS federations will *not* be validated in the sense that their outcomes can be considered predictions of potential real-world outcomes with known accuracy.[2] To be able to meet this strict standard—i.e., determine that a model gives reliable quantitative predictions—a number of comparisons of model output with real data must have been made over a sufficiently large set of cases, and a statistical analysis conducted to determine the accuracy of the model.[3] This is, of course, not feasible with ADS federations that may exist for only short periods of time or include hypothetical systems. Thus, an ADS-based study must carefully consider how a nonpredictive simulation can support the analysis.

Whether or not a combat simulation is predictive, the study must credibly reason with simulation outputs. The determination of how to reason with model outcomes will heavily influence the detailed design sampling plan. Some examples of model analysis strategies include:

---

[1]Scenario requirements include the forces and time period in which the operations will take place, and for how long U.S. forces might be committed. Will U.S. forces *engage* as part of a coalition, as in Iraq, or alone, as in Somalia? The general political situation must be described, including the role that neutrals might play. The location of the combat theater defines the geographical and environmental conditions.

[2]That is, reliable estimates of the errors and biases are available.

[3]See J. S. Hodges and J. A. Dewar, *Is It You or Your Model Talking? A Framework for Model Validation*, RAND, R-4114-A/AF/OSD, 1992.

- *Predictive*: Model outcomes are predictions of real-world outcomes within known accuracies.[4]

- *Plausible*[5] *outcomes*: While not predictive models, outcomes are called plausible if they are consistent with all information that is available and seen as salient to the analysis at hand. Often this is called "face validity." There are several credible ways to reason with models that produce plausible outcomes. If results remain consistent over a wide range of plausible scenarios, then the results have combined strength. In risk adverse situations, the instantiation of a possible disastrous outcome can be decisive. For example, if there is a chance the United States will take 20 percent casualties in a course of action (COA), then one may determine not to take that COA, regardless of average model outcomes. The latter is especially relevant to ADS for it may not require too many samples.

- *Insight (or hypothesis) generation*: The simulations are run only to see how complex interactions play out to gain insight or propose new hypotheses—even though outcomes are not considered predictions.

- *A fortiori (or bounding) argument*: The model is biased in a known direction, which is useful for reasoning in one direction. For example, if a simulation is known to favor system A over system B, and yet the model shows that B outperforms A, then an analysis has credibly shown that system B is the better one. However, if A outperforms B, it may be impossible to untangle whether it results from the system or model biases.

Different types of reasoning require that the experiments provide different types of information. For extensive discussions on credible reasoning with (nonpredictive) combat models see Dewar et al. (1996) and Hodges (1991).

Federation Issues: Determining the Preferred Simulations, Tools, and Data

The high-level design will tentatively specify the simulations and data the ADS federation will use and generate. Determining which simulations, tools, and data (remote or local) are available and best suited to answer the critical questions within time, budget, and resource constraints, and facility availability are critical

---

[4]A close cousin of this strategy is to assume that, while the absolute values of model outcomes are not predictive, differences between two outcomes are predictive. This is particularly enticing when comparing two or more options. This is also a strong statement to make with little evidence to support it.

[5]Defined by Random House unabridged dictionary as "having an appearance of truth or reason; seemingly worthy of approval or acceptance. . . ."

and are not easy to answer.[6] This is an area that the ADS analysis community is just starting to understand. Resolution, maturity, and community acceptance of tools and federations are critical factors. Simulations such as JANUS are well known, and in some facilities well understood by the staff. Fidelity, fidelity balance, data availability, and functional standards must all be considered and agreed upon. Accreditation of the principal simulations is essential. Use of the HLA libraries of federations and models should help with this. Where possible, it should be determined what components can be reused. It is hoped that the community will build up a set of trusted federations whose applicability to classes of problems can be learned over time.

At a high level, this will be based more on past experiences and preliminary investigations. At a detailed design level, interfederation issues must be explicitly considered. Factors that will guide consideration include:

- Component capabilities: Do the various components model the appropriate objects and processes at the required level of detail. The Army Materiel Systems Analysis Activity (AMSAA), in the Anti-Armor Advanced Technical Demonstration (A2ATD) study, established "entrance criteria" for the various components.[7]

- Interoperability: Do we expect the distributed models and objects to interoperate sensibly.[8] This usually will include a comparison of data, entities, algorithms, terrain, visual displays, and human participants. Potentially significant differences (say, in entity or visual resolution) should be addressed. Differences in resolution are an indicator of potential interoperability difficulties.[9]

- Agreement and compliance on data.

- Agreement and compliance on standards and protocols.[10]

- Network bandwidth, reliability, and latency requirements.

- Security requirements at all of the sites and over the network.

---

[6]For a detailed treatment of this subject, see *Analysis & Engineering (A&E) DIS++ Applications Guidelines*, September 12, 1996 (or later) version.

[7]If the criteria are not met at prespecified times, AMSAA suggests that a study might consider postponing some exercises. AMSAA found that criteria involving the capabilities of visual displays presented to human participants, vis-à-vis how the constructive components modeled them, are particularly important in their experiments.

[8]We define "interoperate sensibly" loosely here. It is meant to imply that there will be no unintended biases that will dominate the effects we are trying to measure.

[9]The resolution comparison is facilitated if the data use similar templates and the algorithms specify input data requirements and outputs.

[10]Such as the DIS protocol data units (PDUs).

- Site and personnel responsibilities and assignments.
- Test and exercise schedules.

The explicit consideration of the above federation issues should, if possible, set quantitative criteria to assess the federation (for example, tolerances in simulation or network reliability). If the criteria are not met, steps must be taken to upgrade the components or adjust the analysis. It may be determined that new data or models need to be developed or existing ones modified. If so, resources must be devoted to this and acceptance criteria should be established, or the original requirements lowered.

Here we are talking about design, not test, considerations. Many of the federation issues addressed in the design should be formally tested. The explicit consideration of these issues in the design stage will mitigate problems discovered in formal testing.

Design of Model-Experiment-Model (M-E-M)

We believe virtual ADS-supported studies are best used in conjunction with traditional methods. Others in the community have recognized this too, and a sequential iterative process called Model-Experiment-Model (M-E-M) is commonplace in Army SIMNET studies. The term model typically refers to runs with closed (non-HITL) models. The experiment usually refers to a HITL simulation or live exercise. When implementing the M-E-M approach, extensive stand-alone constructive modeling precedes the experiment and guides the experimental design, data analysis, and perhaps VV&A efforts. Subsequent to the experiment, experimental results are used to calibrate the model. The HITL results can enhance the simulated decisionmaking in the appropriate portions of the constructive simulations, which are then used to develop additional results. In an ADS context, the term experiment might refer to an ADS exercise involving HITL virtual simulators or, possibly, live systems. M-E-M is naturally iterative in nature so that each subsequent experiment or exercise can contribute information that will improve the quality of the constructive as well as the virtual simulators included in the federation design.

As discussed in Section 2, we envision that ADS federations will often be used in concert with traditional constructive models—perhaps in the M-E-M process. The roles that each of the simulation tools will play and their scheduling needs to be specified. The ordering of such activities also needs to be specified. For example, the Focused Dispatch Advanced Warfighting Experiment plan specified three sets of constructive simulation exercises spaced between two virtual SIMNET experiments and a live field exercise.

How Will the Results Be Communicated?

Determining how the results of the (ADS) analysis will be communicated to decisionmakers is an essential early step in the planning. Written reports and/or Vu-graphs are generally available throughout the community. However, real-time movies, videos, simulation graphics, and/or synthetic playback require more specialized equipment such as that found in The Air Force's Theater Battle Arena in the Pentagon or the Military Operations Simulation Facility at RAND. How one plans to communicate results is one of the most important decisions to make and will influence which tools are used, and for some years which facilities. Visualization is a major contribution of ADS[11] and must be carefully planned.

Top-Level Schedule and Run Size Estimates

Realistic but comprehensive schedule milestones must be agreed upon and distributed to the various site managers and exercise analysts. The availability of virtual simulations and constructive models, along with knowledgeable analysts, is crucial, as is available funding. Milestones will include scheduling final planning documents, model changes, tests, rehearsals, operator training, distributed exercises, and analysis products. A draft schedule should be prepared and distributed around the sites, comments sought on the inputs, and a final plan generated. Experience indicates that a large distributed ADS exercise, if most of the models are reused, could be scheduled and run by an experienced management team in as little as six months. This will depend, however, on the complexity of the scenario preparation, the familiarity of the personnel with the constructive models, and the availability of virtual simulators. At this date, the ADS federations are not mature enough for reliable large distributed quick turnaround analyses.

Run size estimates will be a matter of experimental design, resource availability, and judgments concerning the difficulties associated with using live experimental subjects. A general estimate should be made for the high-level design—otherwise, one runs the risk of finding out that there are not enough samples to answer the mail. Estimates are needed for both virtual and

---

[11]A picture may be worth a thousand words, and a real-time playback may be worth a thousand Vu-graphs.

constructive-only runs, as well as those associated with the various sequential stages. Estimates are especially vital for distributed virtual runs because there will typically be few, if any, opportunities to take additional samples.

When we run Monte Carlo simulations, it is not unusual to use a default sample size of 30 or so for cases on which we wish to obtain good estimates. Of course, the sample size required depends upon both the signal one is looking for and the noise obscuring it. The 30 is a general guideline and in most cases is sufficient to ensure that statistical tests based on asymptotic theorems, such as the Central Limit Theorem, are reliable.

When planning for an exercise testing the potential advantage of incorporating the Navy's Cooperative Engagement Concept (CEC) on a U.S. Air Force E-3 Airborne Warning and Control System (AWACS), the vignettes in the plan were to be run "at least three times in order to provide a valid sample size."[12] Unless there is little natural variability in the process, uncommon in combat simulations, a sample size of three is not likely to provide tight confidence-interval estimates, which suggests that the virtual runs should be supplemented—perhaps by traditional methods. This is an area of ADS that needs further, rigorous study.

Personnel Management Assignments

ADS exercises often contain multiple sites and organizations. Organizational assignments enumerate the people, resources, and facilities that the exercise manager will need to successfully conduct the exercise. The *A&E DIS++ Applications and Guidelines* discusses a list of 11 classes of organizational requirements beginning with the user/sponsor and ending with the exercise logistics representative.[13] It must be clear who is responsible for what activities, and since some of the organizations use support contractors, where the funding and contract vehicles will reside.

---

[12] Joint Engagement Technology Study (JETS), Scenario Description.

[13] A similar list is discussed in *Exercise Management & Feedback (EMF)—Rationale*, October 1996.

4. Detailed Design

The detailed design provides the specifics necessary to execute the simulation experiments in accordance with the high-level design. It is the process of determining the specific number and type of simulation experiments, as well as the specific case inputs and outputs to be gathered. For most of the discussion, it is implicitly assumed that the ADS federation will include distributed virtual components. For constructive-only federations, the discussion regarding human participants is, of course, not applicable.

While every analysis has unique features associated with it, there are general principles of experimental design that are worth keeping in mind when generating the detailed plan for an ADS-supported study. Foremost is that the experiments should be objective. If there are biases in the distributed data, models, or participants, they should be addressed—either in the design or the interpretation of results.[1] The design should be robust to small changes in scenario, data, or federation components—i.e., the design may need to include comprehensive sensitivity analysis.[2] Good analysis usually requires that the design include some replication to estimate and formally address chance variation,[3] which is inevitable with virtual federates. Finally, ADS experiments are most efficiently designed if they are intended to adjudicate well-specified quantitative hypotheses.[4]

Federation Issues

The high-level design will have tentatively specified the simulations and data the federation will use and generate. At the detailed design level, a more explicit

---

[1] Identifying potential biases is exacerbated by the distributed nature of ADS.

[2] Of course, for virtual simulations, it will be impossible to take the samples required to perform a comprehensive sensitivity analysis. Nonetheless, model outcomes will help decisionmakers more if they are known to be insensitive to small changes in uncertain input values. Part of a comprehensive study might show that a subset, say the constructive entities, demonstrate robust behavior to small changes in inputs. This is another reason we recommend a balance of ADS and constructive simulations.

[3] Brad Efron, in *An Introduction to the Bootstrap*, Chapman & Hall, 1993, writes "Left to our own devices we are not very good at picking out patterns from a sea of noisy data. To put it another way we are all too good at picking out non-existent patterns that happen to suit our purposes."

[4] Dr. Julian Palmore is quoted by James Redman, *PHALANX*, Vol. 28, No. 2, June 1995, as saying, "don't just model detail because you have the computational horsepower to do it. . . . you must frame a proper null hypothesis and construct an experiment to test the hypothesis—this is the essence of science, the experimental testing of hypothesis grounded in theory."

consideration of interfederation issues should be made. See Section 3 for a list of some important issues.

Design of Experiments and Run Matrix

The design challenge is to decide which experiments to conduct among the astronomical number of potential simulations, federations, data configurations, and scenarios that could be relevant. The number of experiments that can practically be conducted will nearly always be fewer than the number that could be informative or useful, particularly for HITL ADS experiments. A subset of these will actually be performed. The process of selecting that subset is a major aspect of the experimental design.

There is a large literature on the detailed design of experiments (DOE). In this section, we will address some issues that are particularly relevant to ADS experiments. An expanded discussion on how the experimental need determines the class of design best suited for the analysis is contained in Appendix A, which also reviews a few important classes of designs and contains many references to appropriate texts for those needing more information.

Important Design Considerations

Large distributed federations have many opportunities for failure, including network, hardware, and software failures. The design should accommodate the inevitable reduction in reliability (such as network and simulation crashes). The simplest way is to reserve some "rainy day" runs for rerunning failed experiments. If the rainy days are not needed, extra replications can be made or the experiments can be ended early. Analysts should also take proactive steps to "save" runs from minor failures (see "Contingency Playbook" later in Section 4).

There will be random variation associated with most ADS simulations—even those whose primary software components are deterministic. The variation can come from random draws in simulations, interactive human participants, or variable network latencies. The latter two cannot be avoided and are difficult to control. When this is true, it is important that the design allow one to estimate the random variability—the noise obscuring the signal. Without an estimate of the random variability, it will be essentially impossible to parametrically test hypotheses.

In general, the constraints associated with ADS (limited runs due to network and human availability) increase the benefits of more sophisticated designs. For large

HITL ADS experiments, analysts should consider employing advanced designs such as group screening or fractional factorial designs (see Appendix A). Furthermore, the sample sizes necessary for a typical defense simulation, much less an ADS experiment, typically rule out the exclusive use of the above-mentioned designs without some faith that the results are robust and sensibly general.

An example of where a sophisticated design was required to examine all of the factors of interest is in a virtual study at the TACCSF of the impact of potential retrograde methodologies on airborne laser survivability and theater ballistic missile (TBM) defense. For these tests, the human ABL participants control their orbits to keep the plane safe while simultaneously performing their anti-ballistic missile mission. When attacked, these two objectives are in conflict and quite sensitive to human reactions and decisions. The key outcome measures of effectiveness were whether the ABL survived and how many TBMs were destroyed.

The input variables include the number of red attacking aircraft (two values), alternative ABL rules of engagement (two values), two ABL crews, Red air keep-out range (three values), and the timing between the threat aircraft's run at the ABLs and TBM launches (three values). This exercise highlights the combinatorial difficulties associated with human-in-the-loop exercises. Running all combinations of the above variables (a full factorial design) requires 72 samples and does not include samples to estimate variability or additional values for more inferences on the continuous variables. Unfortunately, time and budget constraints limited the sample size to 40 trials. Moreover, there were other variables the analysts would like to have varied but could not because of the limited number of possible samples. These included factors they believed could be critical, such as different threat aircraft, directions of attack, and TBM launch sequences. Our experience is that there are *always* model variables one wishes to vary but cannot because of time and processing constraints.

A sophisticated design allowed the TACCSF analysts to estimate all of the main effects and two-way interactions—and have some extra samples to assess variability. The design used in the test was based on a one-half fraction of the 72 required samples for the full factorial design. This design requires a sample size of 36; therefore, there are four *extra* cases that can be run to provide estimates of (pure) error or serve as make-up opportunities for lost runs. This design is recommended in the National Bureau of Standards Applied Mathematics Series 58: "Fractional Factorial Designs for Experiments with Factors at Two and Three Levels," reprinted in McLean and Anderson (1984). The design is (nearly) orthogonal and allows estimation of all seven main effects (including quadratic for three-level factors) and all 19 first-order interactions, plus random and pure error.

The limited samples also suggest the use of constructive models to identify what factors are critical and the parameter ranges within which results may exhibit significant sensitivities. In this example, some constructive preruns might have better focused the Red air keep-out range and the time interval between the threat aircraft's run at the ABLs and TBM launch variables. More information might have been extracted from the scarce virtual runs. The analysis team would have made constructive preruns if more time and resources had been available.

Making Efficient Designs

With HITL ADS we must squeeze as much information from each sample as possible. Additional design considerations that may be appropriate in simulation with sample-size restrictions are:

- Variance reduction techniques (VRTs): VRTs are a means of reducing the variance of an output (random) variable. We can reduce the samples required to obtain an estimate of a specified accuracy; or, alternatively, we can get more accurate estimates with the same sample size. This may require planning in the design and coding of the simulation. See Law and Kelton (1991).

- Sequential designs: Sometimes a result is clear before all the samples are taken. In such a case it might be efficient to take some of the samples planned for one particular result and use them on another issue. That is, adapt the design as information comes in. This falls under the general rubric of sequential analysis (see Wald (1947) for the classic reference on this topic).

Select Input Data Sources

The distributed nature of ADS puts an added burden on obtaining input data. Ideally, the data (i.e., objects and attributes) will be formally accredited by sources such as the Defense Intelligence Agency (DIA). It is preferred that the data be common to all sites, although for simulations using different algorithms or levels of resolution this generally will not be possible.[5] If not common, the

---

[5]For example, some combat simulations use elaborate algorithms, with many factors considered, to assess the probability that an engaged target is killed, whereas others use simple aggregate table look-ups that are functions of only a few inputs. How simulations treat terrain also varies considerably.

data should be as "consistent" as is reasonably feasible. When the data cannot be made "consistent," they represent a potential source of bias and subsequent steps should be made to either understand the effect or qualify the results. Most of the data should already exist (e.g., system performance and terrain data). Ideally, object definitions in previous or similar exercises should be used.[6] Other data, such as tactics and plans, may need to be developed.

Allocate Experiments to Appropriate Tools

In the ideal analysis we are discussing, the analysis team has a set of tools, including stand-alone constructive simulations and ADS federations. We have argued that each of these tools brings different strengths and weaknesses to an analysis, and that in many situations the best analyses will use combinations of each (see Section 2). It remains to determine which tools, and combinations and sequences of tools, can best obtain the required information. The information needed (e.g., hypotheses to test) determines which tool, or combination of tools, will be used. In some cases, a portion of the information needed may require only virtual or constructive runs, not a combination—although we believe that they will almost always benefit from an interplay. For more on this, see Lucas, Bankes, and Vye (forthcoming).

Primarily Constructive Experiments. In many cases, a large number of runs must be made, to vary the necessary inputs, to get the required precision, or to examine a large number of plausible scenarios. In this situation, the sample-size restrictions on HITL experiments imply that this information *must* be investigated primarily using stand-alone (or single-site) constructive simulations. For not so severe sample requirements, especially in joint analysis, a constructive federation may be appropriate. Of course, all closed models contain many limitations and assumptions regarding human decisionmaking. Therefore, their conclusions can be strengthened by showing that results are consistent with and informed by the results from live and virtual exercises.

Primarily Virtual/Live Experiments. The information needed is often integrally linked to human performance. Examples include human factors (i.e., man-machine interface) studies and command and control evaluations. For these, human participation is often essential, and the experiments will be primarily HITL virtual (or live). Since there are typically few such experiments available, it is preferred that the information obtained (e.g., direct-fire reaction times) occur

---

[6] For federations using the high-level architecture, there will be a community library of accredited objects and models.

repeatedly within each experiment—as is typical of process information. It will be extremely difficult to obtain reliable and precise measures on run-level outcomes (such as kill rates) over a broad range of inputs. Furthermore, the virtual/live experiments should be designed to inform many human process issues (adjudicate several hypotheses). Since the HITL experiments are so small in number and important, it can be valuable to perform an advance constructive experiment to determine the regions of model space where the variables to be examined are the most informative.

Combinations and Sequences of Constructive and Virtual/Live Experiments. Much of the information will come explicitly from combinations of tools. An iterative approach to this is discussed in Section 2.

Effects of Using a Training/Operational Exercise for Analysis

A fact of life is that ADS exercises are expensive and benefit from actual warfighter participation. As a result, many analysis efforts to date piggyback on training/operational exercises (e.g., the series of Army Advanced Warfighting Experiments (AWEs)). This is a good and virtuous thing to do; however, it comes at a high price—which limits the types of analysis it supports. Analysis challenges include a lack of control in the cases that are run and confounding arising from learning effects. These and others are discussed in more detail in Section 10, Effects of Using a Training/Operational Exercise for Analysis. If a portion of a study's experiments are training exercises, provisions must be taken to minimize the analysis challenges, which include supplementing the training exercises with other (perhaps constructive-only) simulations and accounting for participant learning when analyzing simulation outcomes.

Determining Appropriate Human Participants

For most experiments with human participants, special considerations must be taken in selecting the sample or in extending results from the sample to the whole population. Ideally, the humans will be a random sample from the target population, say F-15E pilots.[7] Furthermore, the sample should be large enough to account for differences in the population (i.e., the randomness due to variations in skill level). However, in the real world of combat simulation obtaining a large and representative human sample is rarely, if ever, feasible.

---

[7]Of course, there are cases when one will not want a random sample; when, for example, a sample of the best or worst from the population may be more appropriate. Additionally, all random samples are not created equally. See Cochran (1977) for details on efficient random sampling.

Nonetheless, one should do the best one can and caveat results with respect to shortcomings. For instance, if the F-15E simulator participants are not field pilots, but stationed at the simulator facility, then any conclusions must be qualified with this fact.

Variability among the participants will likely increase the overall variability in outcomes. In statistical terms, the pilots are a random effect (their differences in abilities cause some of the differences in outcomes). As such, testing for differences among pilots and interactions between pilots and other effects can be more difficult. For more on this topic see Montgomery (1991).

Mitigating Learning Effects and the Randomization of Human Participants

It takes many samples to reasonably estimate effects in models containing random elements. Because of the expense and logistics involved with human participation, it is likely that study participants will be used in repeated trials. Any analysis study that wants to generalize the results must therefore be designed with participant learning effects in mind. Say that a virtual ADS-based study, with a fixed set of human participants, is comparing the effectiveness of two systems. It must not take all of the replications of one system prior to the other, because learning among the players will likely bias the results.[8] In this case, the systems' effectiveness would likely be confounded with learning effects.

To mitigate learning effects related to a specific scenario, one may want to randomly and dynamically change the platforms that the human participants play in different replications. For example, one could randomly assign the human participants to elements that used semi-automated forces (SAFORs) in previous runs. A more ambitious approach would attempt to incorporate learning effects explicitly within the analysis; that is, quantitatively estimate and then subtract out the learning effects. Furthermore, if time and funds permit a train-up period may help get participants over the steep part of the learning curve.

Randomizing participants to sides and platforms can also reduce unintended design biases—which is why randomization is a fundamental principle of experimental design (see Fisher, 1947). For example, if the design assigns the best pilots to the Blue side, it has systematically favored Blue—and confounded

---

[8]Learning can take place on a number of dimensions—operational experience, simulation system familiarity, and unrealistic knowledge of the analysis scenarios and consequences of actions taken in previous cases.

outcomes. Randomization reduces the odds of this occurring and systematically favors no side or platform. Of course, if the participants' quality is known in advance they should be assigned (randomly) so that no side or platform has an unintended bias.[9]

Data Analysis Plan

Planning for Data Collection

The distributed nature of ADS and the information typically passed over the network puts an added burden on data collection. Comprehensive, and, if necessary, redundant data collection should be planned as soon as possible after the basic analytic plan has been finalized, a process that experience indicates could take several months and several meetings. For each phase of the exercise, such as Offensive Counter Air, Defensive Counter Air, Theater Ballistic Missile Defense, Ground Combat Operations, and Air Interdiction, each measure of merit (MOM) and related measure of effectiveness (MOE) and measure of performance (MOP) should be listed as a function of which exercise site (or federate) generates which data using which simulations. Data collection must be prioritized rather than collecting everything in sight. The focus should be on collecting data that are necessary to answer specific questions rather than collecting everything and sorting it out later. Advance planning avoids the inadvertent omission of critical data. For crucial data, or sites or simulators of limited reliability, redundant collection may be in order.

We must ensure that we can collect the key outputs, which may include recording audio communications and video collection of information on how humans-in-the-loop are functioning. It is important to remember that we may need to know why things happened, not just that they happened (or did not happen).

Planning for modifications to generate data not currently generated by simulation components (e.g., data probes) is likely to be needed. Current DIS standards are strong on what happened, such as who fired upon whom, but weak on why the firing event happened (e.g., entity perception). For HITL participants, the why information can often be obtained with structured postexercise interviews.

---

[9]In statistical parlance this is referred to as blocking—it is a fundamental principle of experimental design.

Development of controls for dynamically activating/deactivating collection of certain data is important for efficient data collection and analysis. RAND's JANUS model, when run on a sample of 30 vignettes, produces approximately 300 megabytes of data. Clearly, this amount of data would burden the data collection and analysis system if some arrangements were not made to "cull" the results. Provisions for data repositories must be made.

Planning for Data Analysis

Planning for comprehensive data analysis may be more important than a single set of exercises or experiments. As was suggested in the Model-Experiment-Model discussion in Section 3, the federation may be improved in different dimensions from previous exercises, and we should carry those improvements forward into the current exercise. Therefore, the design of the composite database that includes needed data from all sites must be sufficiently user friendly to enable subsequent exercise teams to understand the database procedures. Recent advances in World Wide Web front ends, married to commercial off-the-shelf relational database packages, make this goal more achievable than it once was. This database effort should include documentation, software, and procedures for generating the composite database and using the data contained within the database.

Design and implementation of analysis algorithms, search algorithms, and on-line documentation are essential. This composite database system should also include data transformations for aggregating or reducing the data needed to facilitate analysis supporting the test objectives. If the federation to be used is believed to be sufficiently general that it could support other exercises, thought should be given to incorporating common data aggregation and reduction even though these capabilities may not be required by the existing exercise.

Special consideration should be given to analysis algorithms, that may be simplified by data transformations. For example, the conversion of a PDU database[10] to a database of entity timelines is extremely useful in many analysis contexts. The ability to create an accurate log of the PDUs, replay the PDUs in the correct order, and provide the capability for automated search of the PDUs could greatly simplify the process by which the analyst gains an understanding of what caused the resulting exercise behavior.

---

[10]This also applies to any database in which records correspond to individual events.

Contingency Playbook

For distributed federations it is highly probable that there will be failures of federate elements such as individual models, displays, networks, or even human participants being unavailable. This is simply a function of the large number of potential things that can go wrong compared to traditional analysis, and the immaturity of the technologies.[11] With few samples typically available, each one is precious and should be saved if possible.

Some of the failures can be anticipated—e.g., network or model crashes. For those, a contingency plan should be developed in advance[12] so that given a failure, the exercise director can determine whether to continue, pause, or cancel a run, and how best to accommodate the failure. For example, when a simulation crashes (or a network connection is broken), it must be determined if and how to bring back the entities. In particular, what are the reconnected or reconstructed entities' "situational awareness" and object status?[13] Critical factors in the decision will include the number and types of entities lost, how long they are disconnected, how well they can be reconstituted to their current status,[14] how critical they are to the primary MOEs, and their status in the battle. In some cases, it may be possible to have a lost federate's objects "taken over" by another federate.

The contingency plan should also consider what might happen if manpower or simulation components become unavailable at short notice. These assets could be lost because of budget changes or unanticipated priority conflicts. Contingency plans might include identifying potential component replacements, such as back-up simulations or former active-duty personnel now working as analysts. Furthermore, criteria to postpone an exercise if critical components (e.g., warfighters) are missing should be carefully considered well in advance of the scheduled exercises.

---

[11]Our observations of several large distributed DIS-based exercises indicate that, at present, for any *extended* run one is more likely than not to observe a serious network or model failure.

[12]As opposed to the exercise director trying to make decisions in a high-stress real-time environment.

[13]Object status includes such items as fuel and ordnance. Situational awareness includes factors such as track files and intent.

[14]For example, do we have credible values of their fuel, weapons, and other data? These will typically be available only if periodically current status is saved.

Computer-Generated Forces Considerations

Computer-generated forces (CGFs) are used in a number of ways. Obviously, constructive-only federations rely entirely on CGFs to simulate human participants. In HITL experiments, CGFs can be used in several ways. These include:

- Supplementing the limited number of HITL participants.[15] Large-scale exercises in particular cannot associate human participants with all the scenario's entities. In these cases, CGFs are essential to fill out the scenario.

- Providing standardized responses. In training situations, CGFs can be useful to provide interactions that are tailored to provide specific behaviors, especially doctrinal. In analysis situations, a greater degree of repeatability can be achieved through the use of CGFs, since some of the variability associated with human participants is eliminated.

- Providing deliberately unrealistic capabilities. It may be desirable to particularly stress trainees, or for analysis, to present an unrealistically capable threat to assess extreme cases. While CGFs as implemented today are generally inferior to human participants, they can outperform humans in selected areas. These areas include greater capacity in the face of high workloads, and greater situation awareness (when realism is deliberately violated by providing CGFs with information that would not be available to a human participant in actual combat).

One should be clearly aware of the special features, especially limitations, of CGFs used in experiments. In general, CGFs do not exhibit the flexibility of humans and are distinctly inferior at performing certain tasks, such as explicitly incorporating reasoning about situational uncertainty into their decision processes. When the analysis critically depends on the execution of such tasks, the use of CGFs must be approached with great caution to avoid conclusions being confounded with how the CGFs are modeled. If there is a mixture of CGFs and human participants, the design should be such that conclusions are not confounded with who played whom. For instance, if all Red pilots are CGFs and all Blue pilots are humans in simulators, then exchange ratios are confounded with differences in the Red and Blue pilots.

CGFs need to be calibrated to human performance timelines. Many missions that are the focus of current analytic efforts critically depend on high-speed command and control (C2). An example is attack operations against critical mobile targets.

---

[15] In practice, the majority of decisionmaking entities in HITL simulations are CGFs.

If CGFs are used in these analyses, it will usually be imperative to calibrate the timelines associated with CGF tasks. This is a place where selected HITL calibrating experiments using ADS can play an important role, and should be included in the analysis plan. For a more comprehensive discussion, see Callero et al. (1994).

Design Questionnaire for Virtual Participants

Virtual exercises are important opportunities to get warfighter feedback. A well-designed formal questionnaire helps ensure that the necessary subject data are collected. The questionnaire should supplement and not replace an After Action Review. A formal questionnaire allows quantitative analysis of survey answers—for example, 9 out of 11 pilots felt the system would benefit them in the field. A good questionnaire is short, straightforward, and only asks for critical information. Where possible, the questions should have numeric answers (to facilitate analysis). Warfighter experience should be especially valuable in assessing the validity of tactics, scenarios, and simulations. Of course, unanticipated events and insights are bound to occur. Thus, the questionnaires should provide room for general free-flowing impressions on any topic the participants feel they have something to contribute.

Testing and VV&A Plans

Part of the detailed design includes the planning and scheduling of testing and VV&A activities for the ADS simulation and its components. These topics are briefly discussed in subsequent sections.

5. Exercise Preparation

In what follows it is assumed that at least one distributed virtual exercise will take place. For comprehensiveness, we touch on just a few of the analysis issues here. The EMF rationale document and A&E guideline cover these topics in more detail.[1]

Exercise Roles

Exercise roles, comprise a summary list of the exercise preparation activities that the exercise management will need to perform to successfully conduct the preparation phase of the exercise. The preparation process begins with the user/sponsor, who is responsible for defining the need for the exercise and for securing the resources necessary. The user/sponsor also appoints or selects, as a minimum, a manager for the exercise execution itself, the chief analyst who develops the analysis and data plans, and the VV&A manager. Clear understandings must be developed concerning who is responsible for what preparation activities. Once again, since some organizations use support contractors for these functions, it is important to understand from where the funding will come, and where the contractual vehicles will reside. Both the above-referenced documents provide a list of potential roles and define their interactions. The A&E has more of an emphasis on analysis, while the EMF accents conducting exercises.

Network Issues

The ADS site managers, analysts, and network manager must work together to ensure that sufficient bandwidth exists to support the information flows necessary to run the exercise. In the past, large exercises like the STOW-E had bandwidth problems that were addressed, though not solved, by the use of application gateways to help reduce the network traffic.[2] In addition,

---

[1]For a more complete discussion, see IST, *Exercise Management and Feedback (EMF) Rationale*, October 1996, or *A&E DIS++ Application Guidelines*, September 1996.

[2]Application Gateways (AGs), developed by NRaD, reduced the network flow in the STOW-E by 80 percent. Still, some PDUs were lost. See T. R. Tiernan, K. Boner, C. Keune, and D. Coppock, *Synthetic Theater of War-Europe (STOW-E): Technical Analysis*, Naval Command, Control and Ocean Surveillance Center, May 1995.

technological advances will surely enhance the ability to handle increased bandwidth. However, if recent large exercises are any indication, our appetite for bandwidth will grow at least as fast as our capability. It may be that the scenario, who plays the objects, and what and how often information goes over the net will need to be modified to ensure sufficient throughput. If the federation allows for priorities on the types of information sent on the net, unrecoverable data elements, such as detonations, must receive priority over recoverable ones, such as position updates.

Simulation Modification

Invariably, each analysis has unique requirements that cause one to modify the component simulations. Time and resources must be devoted to this. For estimates of how long it will take and how many people are required, see Boehm (1981). The following comments concerning modification of the simulations apply to both constructive and virtual simulations, although perhaps with differing degrees of applicability.

In general, constructive simulations are easier to modify because there is no physical hardware. Virtual simulations have large software components that are not much different, with regard to ease of modification, than a constructive simulation. However, the virtual simulations have a considerable physical component that for all intents and purposes is fixed over at least the short term. The locations of switches, throttles, sensor displays and the like in simulators are in general representative of the real systems if they exist, or the real system design parameters if the systems are too new to exist.

It is probably the rule, not the exception, that simulations will require modifications to capture desired effects or functionality for the ADS exercise. The authors' combined experience is that seldom, if ever, is a constructive-based analysis performed with an unmodified simulation. To address new issues generally requires new or enhanced functionality. ADS federations, by virtue of their very nature, will require that simulations be modified to make them compatible with each other, and with data collection and processing systems (for example, a data logger). Each of the terms in the phrase "Advanced Distributed Simulation" introduces a need for modification of the simulations intended for use in a federation. In a distributed environment, this may include modifications to make the components compatible, compliant, and interoperable. It may also require instrumenting simulations, say with data probes, to capture data for analysis.

Configuration Management

Configuration management must control the acquisition and modification of models and data to ensure that, as components are modified for specific exercise purposes, everybody has the same algorithms and data. Otherwise, differences may cause errors and bias outcomes.

Pre-Exercise Analysis

In most ADS-based analysis efforts, there will be insufficient time or resources to make many virtual runs. Thus, they are precious and we must get the most from them. Constructive pre-exercise modeling and/or other forms of analysis should be performed to ensure that the distributed virtual runs are as informative as possible. The constructive runs (perhaps with very simple models) are made to identify critical input variables and the ranges where they should be varied in the virtual runs. For example, when attempting to employ armed helicopters against urban guerrillas, it might be useful to use a constructive survivability model to determine if the attack helicopters should attempt to remain above the effective range of shoulder-held surface-to-air missiles (SAMs), or depend on countermeasures and hiding around buildings.

In addition, local runs of virtual simulators that are not connected to a distributed network can give analysts and operators substantial insights into the performance of the virtual simulator and/or the virtual simulator operator. Since full-up distributed network exercises are so expensive, time-consuming, and often fraught with reliability problems, every attempt should be made to learn as much as possible before running the big exercises.

6. Integration and Testing

ADS federations often are synthesized from multiple complex models. The resulting simulation can be extremely complex. Many of the models may have been developed by disparate users for disparate purposes. Developing an understanding of a linked system, including its limitations, to ensure that the exercise will produce meaningful outputs, requires carefully integrating and testing the federation components. Integration and testing is best done in an incremental nature. The team should build from specific events (e.g., detections, engagements) between two components to testing more complex interactions, eventually to the whole scenario. Methods, software, and procedures that automate the process as much as possible should be developed and employed.

There is a natural tension between making this guide comprehensive or complementary to related papers. As discussed earlier, this guide is meant to be complementary with the *Exercise Management and Feedback Recommended Practices* and the *Analysis and Engineering DIS++ Application Guidelines*. Both of those documents have a greater depth of coverage in specifying roles and responsibilities, exercise management, training issues, and, in the latter, engineering applications. We attempt to be comprehensive on the major points while avoiding unnecessary duplication. Our emphasis in this section is on testing the credibility of the information that can be obtained, rather than on the detailed logistics of hardware testing. The most attention is given to interoperability, for that has proven to be one of the most difficult analysis challenges. Furthermore, we will emphasize empirical, rather than theoretical, tests.

Testing Simulation Components

Before attempting to test the whole federation, where possible, individual components should be tested—i.e., verified and validated to some level. Extensive testing should need to be done only for new or modified models and data. See Beizer (1990) for more on testing simulations. It is hoped that in the design most of the components can be taken from existing repositories of well-tested and trusted simulations.

Testing for Compliance, Compatibility, and Interoperability

This subsection covers some (not necessarily all) of the steps one should take in testing a distributed simulation. They have been identified by early practitioners as especially critical—without which expensive exercises may not reach their potential. Comprehensive testing will include checking for

- Compliance: The simulation components comply with agreed-upon standards and protocols, such as DIS 2.03 or the HLA.

- Compatibility: The simulation components function together harmoniously; that is, do they exchange and use relevant data reliably. Agreement in standards and protocols goes a long way here.[1]

- Interoperability: The simulation components are such that the inevitable biases caused by differences in models and data are small relative to the effects of interest.[2]

Ensure Everyone Is Using the Same Enumeration List

A common enumeration list is essential for analysis. We propose, where possible, an automatic executive process check for exercise compliance at the instantiation of a new object type at each site. Manual testing, while better than nothing, is tedious and error prone.

Testing for Data Package (PDU) Accuracy

As with the enumeration lists, all the fields in data packages should be tested. The extent that this needs to be done diminishes as federates mature.

Testing for Data Consistency

Data sets can not be reliably compared without the context of the algorithms in which they are used. If the algorithms are different, the same data may produce

---

[1]We view compatibility as others have viewed interoperability. For example, the Army Modeling and Simulation Master Plan defines interoperability as "The ability of a set of M&S to provide services to and accept services from other M&S and to use the services so exchanged to enable them to operate effectively together."

[2]What is small will depend on individual analytic requirements. There is a subjective element to this and the biases can be hard to estimate.

different results.[3] Experience has indicated that subtle differences in data and algorithms can have substantial biasing effects; see Russo (1996) for some examples.

Detailed Testing of Object Templates. In many cases, there will be instantiations of object classes at multiple sites, e.g., the class of M1A2 tanks. If both sites' objects use the same templates, it will be possible to check that the objects have the same subobjects and attributes, and are defined similarly. Areas that are different should be examined further, for they may cause a bias in the resulting simulation; for example, unrealistically and systematically favoring one site's tanks against the other's tanks. It should be feasible for an executive process to check for object consistency at the instantiation of a new object type at each site. Of course, if the algorithms that use the data are different, then consistency in data definition does not guarantee consistency in simulation behaviors.

Testing of Data in the Context of Different Algorithms. When algorithms are different, it is impossible to do one-to-one comparisons between data sets. However, experience indicates some comparisons can be made. For cases in which the data are of different resolution, one should examine, if possible, whether the aggregated data are consistent with the disaggregated data; that is, would it naturally disaggregate that way.[4] For example, do the timeline distributions (or values) in a disaggregated process sum to the equivalent aggregated distribution (or value). For algorithms that are just different, the data should be examined and, where appropriate, modified to ensure that critical interactions, such as detections, engagement logic, weapon effectiveness, movement logic, and visual display differences for HITL participants, do not result in artificial model-based advantages. History has illustrated that how the models simulate environmental factors can, in itself, make otherwise similar models fail to interoperate.

There will likely be model-based biases if one model's entities have an artificial advantage in sensing, thinking, shooting, moving, or surviving over equivalent entities in another model. In the A2ATD Experiment IV (again see Russo, 1996), they found that by properly tuning key data elements that fed different JANUS and ModSAF algorithms they could substantially reduce intermodel biases without modifying the algorithms themselves.

---

[3]For example, at RAND we found that different terrain intervisibility algorithms produced different simulation outcomes on the same terrain resolution.

[4]One must be careful using models of different resolution. Studies have demonstrated that results can differ simply by the resolution of the model. Hillestad, Owens, and Blumenthal (1993).

Component Pairwise Interaction Testing and Beyond

It is assumed here that all individual models have been tested and are acceptable. The challenge with distributed analysis is to ensure that models built by different people for different purposes interoperate sensibly. It will be difficult to subjectively judge the macro behavior of the whole simulation. Therefore, an incremental series of ever more complicated tests should be carried out—to the extent allowed by budget and time constraints. The subsections below discuss two approaches that complement each other.

Testing Process and Timeline Threads. Any experienced software engineer knows that there is no better way to understand whether (a small set of) computer code is doing what it is supposed to than to trace it through step by step—or even line by line. We propose something analogous, at the appropriate granularity, for critical simulation processes. For example, one could follow an aircraft and see who sees it, when they see it, how they interact with it, engage it, . . ., and vice versa. Graphics are invaluable here. It is important to ensure not only that the data are being passed between the federates, but that the federates are using the data as intended. The flight to be scrutinized should interact with as many key elements as possible. The same flight could be examined as more and more entities and federates are added and tested.

Testing for a Fair Fight Versus Intermodel Biases (Empirical). The key to widespread success with ADS will be the ability of standards and protocols to allow disparate models to interoperate *sensibly*. Combat models are complicated. The idea that standards can make simulations with different algorithms, approaches, or purposes and different levels of resolution interoperate sensibly is foreign to our collective experiences. Inevitable and sometimes subtle differences in data and algorithms can have substantial biasing effects.

Analysis results can be biased if differences in how models simulate similar entities and events result in large differences in outcomes. Here, large must be viewed as relative to other biasing factors and the size of the effects that are being studied. This is often referred to as a "fair fight" between model entities.[5] Ensuring that the intermodel biases are small is necessary, but not sufficient, to guarantee accurate estimates of outcomes. Consistent intramodel biases can also lead to erroneous conclusions that are often more difficult to find. Intermodel consistency can often be checked empirically by playing the elements on the same battlefield and switching which elements the models play, then varying the

---

[5]A fair fight, paraphrased from Foster and Feldman (1996), is achieved if the models and their associated architecture do not give an advantage to any model that does not exist in the real world (an artificial model-based bias).

entities and comparing how model results change. If substantial differences occur simply as a function of which models play certain entities, further examination is necessary to understand why and either correct or account for it.[6]

One of the most comprehensive efforts on model interoperability was conducted in the A2ATD Experiment IV. In this experiment, the oft-used JANUS and ModSAF models simulated a Southwest Asia (SWA) battalion-level armor engagement with supporting artillery and rotary aircraft, with each model playing one side. Initially, JANUS won easily whether it played the Red forces or the Blue. Thus, empirically, there existed an artificial model-based bias in favor of JANUS. Such a bias could conceivably confound analysis results. A substantial effort was undertaken to understand what caused the differences and to adjust the data to reduce the bias.[7] The conclusions of the effort include, "Interoperability between two disparate models is a difficult feat to accomplish, however, it is paramount if analysis is the goal."[8] While there was movement toward interoperability, the improvement could not be guaranteed to generalize to other scenarios.

Consistent with what we have seen elsewhere, the understanding of what caused biases in the A2ATD Experiment IV was achieved by carefully looking at simple small vignettes, finding the causes of model-based differences, modifying the data or algorithms to achieve rough interoperability, retesting the results, and then repeating the steps with other (eventually larger) vignettes. Key factors that caused the biases were detections, acquisitions, and engagements.

Three areas that have proven difficult in achieving interoperability are imbedded decision logic,[9] situational awareness,[10] and environmental effects. If two models handle these factors with different paradigms, it will affect many of the critical events of detecting, acquiring, engaging, and reacting—even if those algorithms are otherwise identical.

We believe that interoperability can be demonstrated only by empirical testing. It will not be achieved through the use of disparate validated models.[11] This

---

[6]What substantial differences are depends on the use and, if the models are stochastic, must be stated statistically.

[7]Of course, data sets can not be reliably compared without the context of the algorithms in which they are used. If the algorithms are different, the same data may produce different results.

[8]Russo (1996).

[9]For example, one simulation, say ModSAF, may use expert-based decision rules, while the other simulation, say JANUS, uses human controllers and scripted movements.

[10]For example, do they track objects and combine disparate information sources (data fusion) with error? Are there errors in identifying whether tracks are hostile? Do they permit fratricide?

[11]Currently there are not many (rigorously) validated models with which one can start the process.

incremental empirical approach is necessary, though not always sufficient, for reliable analysis. We have emphasized an empirical approach because it is relatively straightforward. There are more theoretical approaches to assessing model interoperability. For a rigorous discussion of this, with references, see Harmon (1996).

The Big Challenge: Interoperability. Experiences like the A2ATD Experiment IV have caused us to ponder: are there successful examples of widespread interoperability among separately developed and maintained computer codes? If so, what attributes of these codes allow our simulations to use them without worry? Certainly libraries of mathematical functions contain separately developed and maintained codes with which our simulations regularly interoperate successfully. For example, consider the trigonometric functions, such as Sin. More complicated functions, such as linear programs and even random number generators require more understanding and testing before we trust them—even these well-defined functions do not just "plug and play."

Although the interoperability of such library functions is a great deal less complex than the interoperability of ADS codes, we know of no other more complex example where such complete interoperability has been achieved. We thus believe there are useful lessons to learn about characteristics of codes that interoperate successfully. Four attributes that relate to our ability to easily and reliably use models from a library are well-accepted approaches, documented standards, trusted and tested implementations, and easy accessibility.

How do typical remote ADS models and objects match up with these attributes? Unfortunately, not very well. Different approaches, even resolutions, are used for different problems—there is no one-size-fits-all model. The standards and models are often not well documented. Most implementations require extensive local expertise about known limitations and valid applications—in fact, most models have not been verified and (especially) validated in a rigorous manner. Finally, access to the inner details of models and data belonging to other sites, services, and corporations has not been done well. Attention to these four attributes may help analysts judge potential interoperability of federates being considered.

Assess Impact of Network Effects, Including Latencies

Latencies or network failures may affect critical simulation events, such as air-to-air combat. By artificially generating latencies and/or network failures during tests, one can get a feel for (or rigorously establish) what levels are tolerable. If there is a good chance that the estimated latencies and reliability of the network (which can be estimated) are potentially worrisome, then remedies can be pursued. Potential remedies may include reducing the size of the scenario, cutting the number of sites participating, or modifying the distribution of entities across the distributed sites.

Rehearsal(s)

Experience indicates a steep learning curve. As we have discussed, the best analysis will have several ADS portions spaced in time. The first combined exercise should be more of a dress rehearsal than a generator of reliable information. This, of course, does not mean that if everything works out the results may not be used; just that one should not plan on it. Until federations get more mature it may be that the real analysis runs should not be set in concrete until an acceptable rehearsal has been run.

Testing Data Collection and Analysis

Nothing could be more frustrating than conducting an expensive time-consuming ADS exercise and finding out, after the fact, that the information (i.e., data) generated is insufficient to meet critical objectives. Thus, prior to the ADS exercise, perhaps during a dress rehearsal, a mock analysis process[12] should be constructed to ensure that required distributed data can be successfully collected and synthesized during the exercise. Requirements will include both simulation data logs and other factors, such as the ability to monitor the battle in real time. This will ensure against unanticipated data requirements and confirm that there will be no schedule (or other) problems associated with data reduction and evaluation.

Testing of Exercise Control Mechanisms

During the rehearsal or other distributed testing, the exercise monitoring and control mechanisms should be exercised as fully as possible. Key factors that should be exercised include starting, pausing, and freezing sessions; creating and deleting entities; dynamically accessing critical simulation data (e.g., object status); and exercising the contingency playbook in controlled (and spontaneous) failures. These tests should also be used to determine whether the real-time

---

[12]It will typically not be feasible to do all the steps intended for the analysis; however, an outline could certainly be walked through, with detail provided in the most critical areas.

monitoring capabilities (e.g., visual displays and data probes) are sufficient or should be improved upon.

Security

The tests should exercise the security mechanisms, even if unclassified data are used for the tests. When feasible, the real data should be used so that the test is as close to the real exercise as possible.

Data and Model Enhancement Freeze Date

Large distributed federations can always be improved upon. They can also be unpredictable. Changes to some parts may unintentionally and dramatically affect other, seemingly unrelated, components. Experiences at multiple sites, such as the Theater Air Command and Control Simulation Facility (TACCSF) and AMSAA, has led these organizations to establish firm "good idea" cutoff dates, after which no modifications can be made unless the whole exercise depends on it.

7. Exercise Management

A good deal of effort has been put forth to establish guidelines for the "real-time" management of distributed simulation exercises. This section discusses a few particularly important analysis functions and actions necessary to initialize and carry out the distributed exercise. A more comprehensive treatment of this topic is contained in the proceedings from the series of DIS workshops and in the draft *Exercise Management & Feedback (EMF)—Rationale* (Institute for Simulation and Training, October 1996). The *Rationale* lists the responsibilities of the "Exercise Manager" as "creating, executing, and conducting the exercise. The Exercise Manager is also responsible for post exercise activities." Many of the planning and set-up functions have been covered in previous sections of this report.

Participant Preparation

The exercise initialization can involve the final phase of participant preparation—the delivery of final instructions to human participants. Participant learning effects determine how this function is performed. As mentioned earlier, participant preparation should progress to a point where participant behavior is "consistent" with the real-world environment that is being simulated, but which does not result from participants "learning" the scenario. On one hand, participants who have learned the scenario may "game" it, resulting in widely varying reactions to the exercise situations. On the other hand, actions of participants who know what to expect may display less variation than might be expected otherwise. In either case, if the outcomes of the exercise are influenced by the reactions of human participants, the validity of the exercise will be adversely affected by participants who are over- or underprepared.

The problem of how much to prepare exercise participants is less critical in simulations where the human-in-the-loop (HITL) interactions take place in a simulation environment that closely resembles "real-world" conditions. For this reason, simulator fidelity, as it continues to improve, will not only allow the exercise to mimic more closely real battle conditions, but it will also reduce the level of participant preparation, thereby eliminating the corresponding problems associated with over- or underpreparation.

Network Monitoring

Effective network monitoring will contribute to the proper transfer of information and the integrated, continuous participation of exercise entities. A first step in network monitoring would be a network initialization carried out by all entities. At intervals during the exercise, similar "check-ins" can be carried out to check that each exercise entity is functioning properly. Periodic checks of the network save the effort that would be needed to check all entities at the shorter time steps at which PDUs are being created to carry out the exercise. Real-time monitoring may also allow problems to be solved before they threaten the exercise as a whole.

The effectiveness of the Exercise Manager may also be critically dependent on network monitoring. For large distributed exercises, Stender (1996) suggests that exercise objectives may be best served by continuing a simulation even if one or more entities experience networking problems. Applications or entities could even be removed from the network in an attempt to solve a potential problem. Such actions, although seemingly drastic, might be the best course of action in a large exercise, in which the costs of stopping the exercise, in time or money, may be higher than the costs of running without full participation. To make such an exercise management plan work, specific ground rules, determined in advance, are necessary. As the enforcer of the exercise ground rules, the Exercise Manager would need accurate and timely information on network capabilities.

Security

The complexity of distributed simulation exercises will often require that security issues be considered from early exercise design phases on through postexercise analysis. The EMF *Rationale* discusses the responsibilities of an "Exercise Security Officer." In addition to the security of PDUs, other forms of communication and information must also be secure because they may involve sensitive or classified information.

Current bandwidth requirements often result in the use of dedicated lines to link the entities in a distributed simulation exercise. Combined with current encryption technologies, these systems provide security levels sufficient for many exercises. Other arrangements can be made for simulations involving entities with higher security requirements. While the central data-gathering and simulation oversight facility would generally need security levels equal to the highest of any of the simulation entities (although exceptions to this requirement could exist), all entities need not be at that level of security clearance. Traffic management for PDUs could follow a set of "need-to-know" rules for various entities. Such multilevel security arrangements for distributed simulations have been discussed. As the speed and bandwidth of network technology improve, many safeguards that are now thought to be too costly or time-consuming may become viable options. Smith and Winkler (1996) review a variety of security mechanisms for use in modeling and simulation repositories.

Time Synchronization

For exercises with many entities distributed over wide areas, an absolute time stamp will likely be required for exercise synchronization. The Global Positioning System (GPS) provides an absolute time signal sufficient to synchronize the various entities of a large distributed simulation. Including time-stamp information in the PDUs produced by the exercise entities can serve several purposes. First, this information can provide information to ongoing network monitoring. Second, in case of an interruption in the exercise or for planned breaks between exercise segments, the information can be used to "rewind" the simulation to a particular point in time and accurately restart the exercise. Third, accurate time identification is critical to reconstruction of the simulation as part of postexercise simulation and analysis.

Although a good deal of recent work covers the problems of time synchronization in distributed exercises,[1] Myjak (1966) notes that the use of absolute time stamps has not seen wide application in distributed simulations. The use of absolute time stamps was a primary recommendation resulting from earlier RAND research in which the study team participated in several distributed exercises.[2]

Exercise Oversight

The *Rationale* (Institute for Simulation and Training, Draft, 1996) proposes an "Exercise Manager" who would have "overall technical and managerial responsibility for creating, executing, and conducting the exercise." This person would retain the highest level of responsibility during the conduct of the exercise and would perform several functions.

One function of the Exercise Manager in controlling and monitoring the overall progress of the exercise is to propagate, monitor, and enforce a set of ground

---

[1]For recent papers see Cox et al. (1996) and Myjak (1996).
[2]Kerchner, Friel, and Lucas (1996).

rules. Such ground rules, as discussed earlier, can serve a number of functions, including controlling entity participation, determining ends of exercise segments, oversight of security and network monitoring, and other critical functions. The ground rules will prove effective only for the specific contingencies they were designed to address. For many actions that need to be taken during an exercise, such as reactions to information on network viability, it may be best to have automated enforcement of the exercise rules. Other necessary actions may not require automated enforcement, and indeed the cost of designing an automated system of ground rules for such cases may outweigh the benefits. Other cases involving unanticipated contingencies may arise for which the Exercise Manager will be required to make critical real-time decisions.

In addition to enforcing the ground rules, the Exercise Manager should oversee network monitoring. As mentioned, there may be significant interaction between the results of network monitoring and the ability of the Exercise Manager to carry out oversight functions. The Exercise Manager should also keep a log of the exercise, including problems, interesting events, and observations.

Gathering Data on Human Participants

While the emphasis on gathering data on human subjects might be central to ADS exercises designed for training purposes, it is also important in simulations carried out for analysis. Accurate data are necessary for reconstructing the simulation to include the non-PDU information that characterizes much of the input of human subjects into the simulation. Human participation information and actions must not only be accurately synchronized with PDU-based information, but should include as much content as could reasonably affect the reconstruction and interpretation of simulation outcomes. Human data recording may include audio, visual, and written observer and self-reports.

The second purpose of gathering extensive data on human subjects is to analyze the impact of human inputs on the simulation. This analysis will aid in determining whether human participation was incorporated into the simulation as intended and whether the human inputs accurately reflect the "real-life" conditions they were intended to model. This verification and validation of HITL inputs can also be used to inform and improve the participant preparation process.

8. Postexercise

Immediate After Action Review

After an ADS exercise is completed, the analysis team needs to conduct an after action review (AAR) to capture information about the exercise that is only in the heads of the participants. These data, and the data that were collected during the exercise, are then processed to support the analysis strategy.

- **Identify potential uncontrolled causal (confounding) factors.** Analysts who have primarily used only constructive models should keep in mind that an HITL ADS experiment is not as well controlled as a constructive experiment, and that explicit steps to understand confounding factors are thus necessary. AARs assist in the identification of unusual aspects to a just-completed run.

- **Collection and synthesis of participant questionnaires** will be useful for things like the exploratory ADS runs to identify important human performance factors.

- **Correlation and synthesis of distributed insights** refers to a less formal process than questionnaires, perhaps by videoconference. It can be particularly important as part of the analysis process substep "Determine human performance factors that have important influences on the scenario," part of the step "calibrate/validate human performance factors in constructive models" that we recommend for many analyses. Capturing insights that have to do with broader cause-and-effect drivers in the simulated scenarios is another important function of an AAR. These insights, particularly when they serve to stimulate system or concept of operations changes, are at least as valuable as the quantitative measures derived from simulation exercises. Discussions between participants at an AAR should be an excellent vehicle for eliciting such insights.

- **Synthesize and review trouble reports and needed improvements** is another activity that partially takes place at an AAR. While many of the problems will fall under the purview of the simulation support team, rather than that of the analysts, other problems, such as those that indicate problems or limitations associated with simulation components, will affect the analysis proper. Effectiveness will rely critically on formally logging trouble reports during an exercise.

Data Analysis

The data analysis process for ADS exercises is more difficult than the process using stand-alone constructive simulations, in two ways. First, when HITL is involved all data may not be automatically captured—some data will typically be in the form of forms or notes filled out by observers during the exercise. Additional data of this type can be generated during the AARs. Second, the data that are collected digitally will usually reside in distributed and distinct databases, often at geographically separate locations. These differences from stand-alone constructive simulations imply the need for different analysis procedures.

On the other hand, ADS has features that can benefit the data analysis process, including a high degree of standardization in the form of the outputs available from exercises and the resulting availability of tools that support data reduction and analysis. Perhaps the most important of such standardized tools are visualization tools that can provide a highly flexible capability to observe the simulated exercises, and to display illustrative segments of the exercises for presentation purposes.

Consolidation of Diverse Data Sources

Data analysis will require the consolidation of data collected separately for each federate, and also data digitized from manually collected sources, into a form directly accessible by whatever data analysis tools are used in the effort. ADS will normally provide for the global collection of information necessary for the interaction of federates, such as physical state and communicated messages, but internal state information will not normally appear on the network and must be collected separately.

This point warrants additional discussion. ADS in the form of the high-level architecture (HLA) and by implication its implementation as DIS++, provides for federates to declare their intent to publish arbitrary information as attributes that could be collected by other federates. Thus there is nothing *in principle* to prevent all information needed for an analysis to be centrally collected at the exercise. However, there are pragmatic reasons why this is unlikely to be practical.

The first reason is that network bandwidth is limited: data that are not needed at the exercise to make the simulation operate properly will usually need to be kept off the network. These data could be collected locally by each federate. This can be done efficiently if the infrastructure is capable of filtering information not required by remote federates; such information is not put on (to congest) the wide area network (WAN). After an exercise is over and bandwidth is no longer at a premium, the data could be centrally consolidated. This process would be simplified by the (presumed) fact that the local databases are already in a federation-compatible format.

However, it is unlikely that the process will be as simple as consolidating databases that are already in an ADS-standard format. In essentially every significant analytic effort, it becomes necessary to define customized measures that are specific to the problem. Inevitably, there are new data that need to be extracted from the simulations that are not on the federation's list of data available for publication. Modifying this list may entail making global changes in many federates, so the most efficient way to implement the extension may be to have the federate output the information to a local database that is in a nonstandard format. This approach will reduce the reuse opportunities, both in the generation of the local database itself and in the consolidation software.

Format of the Consolidated Data. The format of the consolidated data will probably not be in the form in which the data are initially stored, even aside from the differences caused by consolidation per se. The data collection process for a simulation of necessity takes the form of recording data items that are generated chronologically. Thus, the data about a simulation entity will not be together, but must be postprocessed into a form that brings relevant data about each entity together. Data about things that do not correspond to physical entities can be even more difficult to collect. For example, an air-to-air combat missile engagement is a process that is distributed over both entities and time. The computation of measures from a missile engagement—for example, the kill rate associated with particular attacker/target types, subcategorized by shot range or angle—entails correlating data from events that occur at different times (shot and endgame), are associated with different entities (attacker, target, and missile), and were adjudicated at different sites.

This processing will clearly be more efficient if the correlation is not performed on the fly, but instead can work from something like missile engagement records. Efficiency accrues because it is likely that slightly different versions of an analysis program will be run repeatedly on the same database, not just once. Probably more important, experience shows that the development of analysis programs is enormously simplified when intermediate records, such as missile engagement records, are utilized. This simplification reduces development time, facilitates modification, and reduces errors.

The ubiquitous use of ADS may be accompanied by a shift to sophisticated commercial database engines, both relational and object-oriented. However, current practice in the analysis community often employs special-purpose analysis codes that use conventional databases (such as DIS logger files) as input. It is appropriate, then, to suggest how to develop these special-purpose codes in an ADS context. We will confine ourselves to a few general suggestions.

Data Analysis Tools

At present, consider using AWK or PERL for a major part of the processing.[1] These languages have great capabilities for processing text, so scripts for analysis with simple text-based records can be developed very rapidly. Equally important, they can be easily modified in response to changing analysis requirements. If formality or additional speed is required, it is easy to convert an AWK program into C or C++, particularly if an associative array library (or C++ class) is available. We will confine comments to AWK, which we are more familiar with, although PERL should be viewed as a more capable successor to AWK and is probably the preferred tool.

The simple text-based records appropriate for use by AWK might include the missile engagement records used in the example above. These records need to be generated from logger files using a two-step process. The first step uses a conventional programming language like C to convert the binary logger file into a text-based form. The second step uses an AWK script to process the text files and convert them into the record forms more suitable for analysis. A key reason why AWK is highly suitable for such processing is its associative array feature. Basically, there is an associative array for each datum type, tagged with an array index (like a missile number) and stored in memory as the primary text file is read. After reading the file is complete, a postprocessing step iterates over all index values and writes the individual arrays into a single record per index value.

If C was used instead of AWK, one would instead define a structure for the target records, and store pointers to each instance of the structure in something like a hash table accessed by the missile number (missile numbers tend to be multidigit and unsuitable for direct use as sequential indexes). This is perfectly feasible, but will be more cumbersome to write and maintain.

---

[1]PERL and AWK are programming languages. The preface to *Programming for PERL*, by Wall, Christiansen, and Schwartz, (1996) describes PERL as originally being "a data reduction language, designed for navigating files, scanning large amounts of text, creating and obtaining dynamic data, and printing easily formatted reports based on that data." PERL has grown much larger, but the above language features are those of primary interest for analysis using simulations. AWK predates PERL, and can be viewed as a useful subset of that language.

Perhaps more important than language choice is an admonition to avoid analysis programs that are opaque and inflexible in the selection of what they analyze. We have seen a number of programs that present fancy window interfaces to users, but that are in fact unusable because they cannot be easily altered to respond to changing analysis processing requirements. It is essential to keep in mind that these changes are normal and frequent in analysis, and that the tools used must support easy modification.

Central Consolidation of HLA Local Databases. The real requirement in data consolidation is for a process, not a single tool. There are multiple steps in consolidating data, and there are ample opportunities to simplify the steps, generally through partial automation. The use of automated steps will reduce the burden, but more important, it will reduce the chance of error, such as selecting the wrong local database for use in the consolidation process. We next discuss these steps and automation opportunities.

1. The first process step begins at the start of an exercise, where as part of the initialization process the identities and location of the input files used by each federate are recorded (centrally, or locally at a location made known to any remote analysis site). This is important because it is often necessary to determine input settings during the analysis process, including obscure settings that would not routinely be preserved apart from the local input files.

   A corollary is that the local input files should not be modified and reused for subsequent exercise runs; new versions need to be used. Less extreme approaches are possible, such as using only a new copy when an actual change occurs, or preserving differences, but it is essential to adhere to the principle that it should be possible to reconstruct an old input. Additionally, this reconstruction should be performable by an automated tool; this would not be possible, for instance, if the file locations were recorded only on a printout.

   Also, at exercise initialization, the identity and location of local output files should be recorded. Furthermore, the initialization software should be capable of preventing the use of already-existing files, which would then be overwritten. For example, the file names could include a unique exercise run ID.

2. The actual consolidation process will entail accessing the various run output files and databases, transmitting relevant contents to a central analysis site, and translating the contents into a standard form that is accessible by analysis software. It is certainly possible that instead of a single central site, a

distributed database management system (DBMS) will be used, but in some sense this is a detail: the point is that the data must be converted into a form accessible by a single tool, distributed or otherwise. Ideally, the transmission of data during the consolidation process would use the same network as the exercise itself, with data centralized during off-peak hours.

Preparation of the tools to perform this consolidation should be undertaken well in advance of the ADS exercise. In our experience, nasty surprises associated with reading databases/files provided by others are frequent, and days or even weeks can be lost if the software to do so is not started early enough.

3. Preprocessing of data, prior to processing that generates analysis measures and other statistics, is generally desirable, as discussed above in "Format of the Consolidated Data." The preprocessing may be effectively performed as part of the consolidation process, although when the development of entity timelines requires use of data from multiple sources, this step needs to be postponed until a consolidated "basic" database is generated.

Quick Data Summary and Review

A quick preliminary analysis is extremely useful to have when going into the AAR. It can take the form of a standardized report generated from unconsolidated data, especially the ADS equivalent of a DIS PDU logger. Visualization tools, described next, are also part of the quick data summary and review process, although their use, of course, transcends just this application. The authors have taken part in constructive simulation study efforts where for all intents and purposes the *only* quick review tool really needed was a visual playback of the scenarios just run.[2]

Visualization Tools

Displays called stealths are used by analysts and others to passively observe distributed engagements. Stealths can equally well be used to examine ADS exercises after the fact, using stored logger data to drive them. Below we describe a few attributes that we think stealths should have to meet analysis needs. We are not aware of any single stealth possessing all of these capabilities.

---

[2]Of course, this is not usually the case. Typically, quick reviews should include examining important summary statistics, such as killer victim scoreboards, critical events by time and range, and so forth.

Viewpoints. Stealths should support a variety of viewpoints or perspectives. To understand how a scenario develops, we would like to have both the macro view, such as overall positions, and the micro information, such as a platform's loadout and situational awareness, the latter sometimes being required for detailed understanding. It would be desirable if stealths allowed users to freely switch views between an overhead view, a view from the cockpit, or the viewpoint of an arbitrarily placed observer.

Finding particular platforms of interest in a very large scenario can be difficult with any kind of view. The ability to have designated platforms change color, or blink, would be extremely helpful, and appears easy to implement. Similarly, the ability to have "dead" platforms, or other critical events, show up with a special symbology would often be valuable.

Extraction of Detailed Information. The small number of runs available from ADS/HITL exercises implies that much of the analysis benefit will come from a detailed understanding of cause-and-effect relationships within a given run (battle). Therefore, improvements to stealth capabilities to provide the kind of detailed information analysts need should be of great value to ADS-based analysis efforts. Such improvements include the ability to designate an entity and have a pop-up window provide detailed information about the status of that entity. For computer-generated forces, this information should include current plans and intent.

Intent and plan information can be partially obtained for human participants by listening to their voice communications. Good situational awareness from a stealth is impossible without access to the communication that is taking place. It would be invaluable to be able to designate a player and then listen in on his or her communication channel, or to be able to select an arbitrary communication channel to listen in on. To do this post-exercise, these communications need to have been recorded, and furthermore automatically synchronized with other logged information during playback using a stealth. The development of this kind of capability is impractical for single analytic efforts. A major benefit of ADS is the promise that such tools, once developed, will be broadly usable for many efforts.

Finally, the analyst will often be interested in particular types of interactions that occur relatively infrequently. It is fatiguing and error-prone to have to constantly watch a screen for events of interest. Thus, one wants the capability to be alerted when a critical event like a theater ballistic missile launch occurs, or when a strike mission is approaching its target.

Model-Experiment-Model (Postexercise Modeling)

The material covering this subject is addressed in Section 2, "ADS Within a Broader Research Plan." Constructive modeling can follow each category of an ADS exercise that occurs in the context of an overall analysis strategy, but perhaps the most important kind is the constructive modeling used to extend the results of an ADS exercise. This is discussed in Section 2, "Extend ADS Results With Additional Constructive Runs."

9. Verification, Validation, and Accreditation

It is generally recognized that formally evaluating models in a verification, validation, and accreditation (VV&A) process enhances the reliability of model results and increases the confidence of model users and consumers. Accordingly, there is a considerable body of research on VV&A, particularly validation. Unfortunately, the history of validation of stand-alone constructive combat simulations contains few success stories. The distributed nature of ADS can greatly complicate all aspects of VV&A. For example, most distributed federations will exist in the desired state for only a short time. Thus, there will be insufficient time to rigorously complete a VV&A process.

The body of work on military models VV&A is considerable and we will touch on just a few philosophical issues and special VV&A challenges posed by distributed simulations. References are provided for those seeking more information.

Definitions of Validation

While all aspects of VV&A are important, validation has proven to be the most difficult. Verification relates to ensuring that the code does what it is supposed to do. There are many software development texts and tools for guidance. Likewise, accreditation relates to the sponsor accrediting the model for his/her purposes and is highly subjective. Validation, on the other hand, relates the model to potential real-world outcomes, as seen in the following definition from Department of Defense Directive 5000.59:

> The process of determining the degree to which a model is an accurate representation of the real world from the perspective of the *intended uses* of the model.

This definition emphasizes validation with respect to use. It suggests you validate "model uses" rather than models. This is the idea behind "Credible Uses" (CU) and is discussed next.

Credible Uses and the Validation of Model Uses

Modern definitions specify that validation is defined for a given use. In practice, the idea of use in validation is all but ignored. In part, this is a result of what people mean by "use." The approach developed by Dewar et al. (1996) concentrates on *how* one uses the model to support reasoning rather than *what* the model is used for. This idea relates to what we called the analysis strategy in high-level design in Section 3. This approach distinguishes the "how" as the logical use and the "what" as the functional use.

Dewar et al. define general classes of "logical" use, such as predictive, plausible, nonpredictive, so that

- most any role of the simulation in any reasonable use would appear in some class, and

- credibility criteria (read validity) can be identified for each class.

The bottom line with most ADS federations is that you validate nonpredictive analysis strategies that use models rather than validate the model itself as a predictor of potential real-world outcomes. Most ADS federations will not exist long enough to take samples over a broad enough set of conditions to estimate how close to the real world they are for essentially all intended uses.

Recommended Practices and Validation

The need to establish standards for DIS VV&A processes was recognized at the 8th DIS Workshop. Subsequently, a special DIS VV&A workgroup was established. Their current VV&A approach (for our outline, see the Institute for Simulation and Training (1996)) involves a nine-step process that emphasizes a comprehensive and high-level VV&A process. It is a more traditional approach to VV&A.

The current first phase of the recommended VV&A process coincides with the development of an exercise plan and requirements. Phase 2 requires that compliance standards be verified for all proposed exercise components. Phases 3 through 7 directly coincide with the five phases involved in "Design, Construct, and Test the Exercise." All of the validation and verification steps involve feedback with the corresponding exercise activity. The phases include: Validation of the Conceptual Model; Verification of the Preliminary Design; Verification of the Detailed Design; Compatibility Verification; and Validation of the Overall Exercise. Phase 8, Exercise Accreditation, requires input from previous phases and postexercise activities. Reports are prepared during all phases of the VV&A process, resulting in a V&V Report and an Accreditation Report. To date, we know of no examples of this kind of validation in a distributed simulation environment.

Practical Levels of Validation

The practice of simulation validation falls far short of what might be envisioned from the definitions above. Pace (1997) describes three "fundamental levels of simulation validation." And, "most (real-world) simulation validations correspond to one of these three levels." Briefly, the three validation levels obtained in practice that Dr. Pace defines, in order of increasing cost, are:

- Inspection-level validation. Here, the validity of the simulation is given by experts who subjectively judge that "the simulation responses" are as one would "expect" real outcomes to be. This is akin to "face validation" and in many cases is the best you can do. That is, constraints in resources and data are such that quantitative comparisons to real data cannot be made.

- Review-level validation. This type of validation establishes that the simulation has "acceptable predictive behavior for (a few) test cases considered." Additionally, the model may have been shown to be "well-behaved" from replication to replication and is "mathematically stable."

- Demonstration-level validation. This validation level is the strongest and requires that "simulation performance and response can be predicted to perform correctly consistently according to objective criteria. . . ." This is feasible only in a few simulations, and requires adequate data to be available over a sufficient range of cases. Furthermore, it is expensive and time consuming to make (sometimes generate) all the required comparisons between the simulation and the data.

Human-in-the-Loop, Computer-Generated Forces, and a "Fair Fight"

With the improvement of DIS and ADS simulation capabilities and complexity, the process of model validation will necessarily become more difficult. HITL components, computer-generated forces (CGFs), and the goal of modeling a "fair fight" provide one example of the problems involved in the validation process for ADS and DIS simulations.

ADS exercises carried out for training or analysis will often involve the participation of HITL participants who can serve a useful validation function. HITL participants offer "built-in" reality checks to ensure that real-world systems are represented accurately and that exercise outcomes, within the range of outcomes allowed by the exercise, seem "reasonable." Validation by HITL participants is known as "face validation." Such determination is important in determining to what extent a particular simulation event—a conflict or other action—is represented "fairly." The ability of HITL participants to support validation efforts can be affected by several factors. First, the number of components interacting to yield particular outcomes may make it difficult for HITL participants to understand how well particular components or events are represented. Second, many human decisionmakers may be represented by CGFs instead of HITL participants. Third, the ADS exercise may be designed to analyze situations that are anomalous or even outside the range of experience of HITL participants.

Parting Thoughts on Validation

We believe that formal VV&A efforts can greatly enhance the analysis products of models. The brevity of coverage in this report should not be interpreted as suggesting that we believe VV&A is not important. It is important. It is simply that the complexity of issues is well beyond the scope here to provide more than a cursory discussion of model validation. We encourage readers to look at the following spectrum of VV&A documents: Dewar et al. (1996), Hodges and Dewar (1992), Defense Modeling and Simulation Office (1996), Davis (1992), Institute for Simulation and Training (1995), Pace (1997), Sanders and Miller (1996), and the special *PHALANX* issue, edited by Palmore (June 1997). These documents also contain many excellent additional references.

Our strong recommendation is to concentrate on validating "uses of models" as opposed to validating the models themselves. This is why we specify the analysis strategy in the high-level design. Our reasoning follows from the belief that it will usually be impossible to compare a distributed federation to real data over a sufficiently large set of initial conditions to view model outcomes as reliable predictions of potential real-world outcomes.

10. Effects of Using a Training/Operational Exercise for Analysis

ADS exercises are (currently and for the foreseeable future) expensive and benefit from actual warfighter participation. As a result, many analysis efforts piggyback on training/operational exercises (e.g., the series of Army Advanced Warfighting Experiments (AWEs)). This is a good and virtuous thing to do; however, it often comes at a high price—a reduction in the ability to design and control cases. Most analysis experiments must be carefully designed and tightly controlled to achieve reliable results. Because design and control are usually severely restricted (from the analyst's viewpoint) in a training exercise, there is an inevitable reduction in the types of analyses that can reliably be supported in conjunction with training.

Simply put, training/operational and analysis objectives can be at odds. Table 10.1 summarizes some of the potentially conflicting aspects between training and analysis exercises. If a portion of a study's experiments are training exercises provisions must be taken to minimize the analysis challenges—perhaps supplementing the training exercises with other (maybe constructive only) simulations and incorporating a learning factor when analyzing simulation outcomes.

Table 10.1

Training/Operational Exercises and Analysis Can Be at Odds

	Training/Operation	Analysis
Primary objective	Attain better skills	Better decisions
Primary beneficiary	Soldier	Decisionmaker
Credibility criteria for simulation	Proper stimulus to soldiers	Information obtainable
Interreplication effects	Learning	Independent
What is typically varied	Current doctrine/tactics and approved scenario (few in number)	Multiple doctrine/tactics in hypothetical scenarios
Typical measure	Qualitative and non-attribute	Quantitative and attributable (cause and effect)

Fundamentally, training is primarily for the soldier, whereas analysis is for the decisionmaker. The credibility criterion for simulations used in training is: Do they provide the proper stimulus for the soldier to learn better skills? If so, they meet their objective, even if the resultant outcomes are highly unrealistic. The credibility criteria for analysis, by contrast, relate directly to the capacity of the simulations to provide information about potential real events. Effective training often provides immediate feedback for errors—for example, having an unrealistically strong opposing force (OPFOR) punish soldiers for tactical errors. This has the consequence of biasing some analytic measures. For analysts studying training exercises, one must always be aware of the adage, "Don't take tactical lessons from training."

There are other fundamental differences between training/operational exercises and analysis experiments. In training, the objective is to improve skills, so learning should occur between replications. This learning complicates analytic comparisons among replications. Additionally, training often involves repetition with approved doctrine. Conversely, optimizing force organization and tactics, for example, requires examining multiple variations.

This is not to say that it is impossible to obtain good analysis in training events. Indeed, there is information that is best gleaned through such events. However, the constraints imposed by training/operational exercises restrict the classes of effective analysis, often to micro issues and those amenable to few event analyses. Classes of analysis that have been credible in previous training/operational exercises include studies of human factors, measurements of process-oriented information (such as delays in transferring information),[1] lessons learned from implementation problems (such as equipment or software failure), qualitative insights obtained from the pattern of outcomes, and establishing that a particular outcome is possible.

Classes of analysis that are difficult to perform in training exercises include comparisons of the battlefield effectiveness of various systems, organizations, doctrine, tactics, measurements of lethality, survivability, and tempo; statistically valid quantitative outcomes; comprehensive sensitivity analyses; and exploratory modeling. Unfortunately, some of these are expected from analysis efforts that piggyback on training events. Supplementing the information gleaned in training exercises with constructive simulations may help compensate for these difficulties.

---

[1]For process-oriented information, a few large-scale events may contain thousands of subevent process experiments.

11. Conclusion

The intent of this report is to serve as a guide to those considering using ADS for analysis. In addition, there are general policy-related conclusions that follow from our findings that deserve discussion. First, we believe that ADS has great potential for increasing the effectiveness, scope, and depth of analysis. However, realizing that potential is more difficult and more expensive than many appreciate.

We can move toward the analysis potential of ADS if the role of ADS in an analysis is carefully specified. In combination with traditional methods, ADS can more credibly represent human interactions and improve this critical component of our models, whereas traditional methods can examine a greater breadth of cases and focus on those conditions where ADS methods are essential. These benefits will not be gained without overcoming a variety of technical, operational, and administrative challenges. In particular, we feel that resolving problems with interoperability among models is essential. Unfortunately, "plug and play" interoperability has not been successfully addressed in contexts that are much simpler than distributed combat simulation. Thus, there is little reason to expect that these challenges can be successfully solved for general ADS combat analysis purposes in the near future.

To improve model interoperability, we need to establish well-accepted approaches to representing combat elements, document the models and standards used, build up trusted and tested implementations through frequent and wide use of the models, and provide easy accessibility to the models. Further research in these areas is needed if ADS is to become an oft-used and credible vehicle for analysis. Finally, given that we believe ADS is often best used in conjunction with stand-alone constructive simulations, investments must also be made in these models and the analysis methods that use them[1]—and the interaction between the two.

---

[1]For example, the exploratory modeling work of Bankes. For an overview, see Bankes (1993).

Appendix

A. Expanded Discussion on Design of Experiments

To inform a critical decision, gain insight, or achieve the purpose of the analysis, information is needed from experiments.[1] The information needs drive both the outputs (MOEs/MOPs) to be gathered and the inputs to be varied—including potentially confounding factors. Of course, not all inputs need to be specified before all the runs begin. As learning occurs, or at natural breakpoints, the run matrix can be modified.

The design challenge is to decide which experiments to conduct among the astronomical number of potential models, data configurations, and scenarios that could be relevant. The number of experiments that can be conducted will nearly always be fewer than the number that could be informative or useful. Figure A.1 illustrates this situation. The hypotheses and assumptions in an analysis, together with available data and constraints on available simulation capabilities, define an ensemble of simulation experiments that might be conducted in support of a given analysis. A subset of these will actually be performed. The number of ADS/HITL experiments will almost surely be only a small fraction of those that would be informative. The process of selecting that subset is a major aspect of the experimental design and is driven by the information needed from the experiments and the information obtainable. The information needed is dependent on how the simulations are being used (see the analysis strategy discussed in Section 3). The information obtainable relies critically on what can be estimated and tested in a highly constrained number of runs.

Specify Detailed Experimental Need

The high-level design has established what general information is required from the experiments. To facilitate creating the run matrix the needed information must be made as precise as possible. General statements of critical operational issues leave too much imprecision. Examples of precise statements, internal to a simulation,[2] which should encompass most analysis studies, include:

---

[1] See Dewar et al. (1996) for a discussion of one approach for determining what information is needed from experiments.

[2] Or across multiple simulations.

78

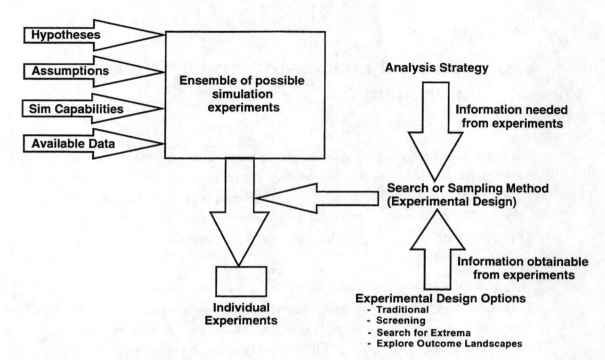

Experiments are derived from an ensemble of possible experiments depending on the experimental need and information that can be obtained.

Figure A.1—Simulation Experiments

- Parameters, effects, and interactions to be estimated. They should include desired confidence levels associated with the estimates (for example, the expected exchange ratios as systems and tactics are varied, with confidence intervals no larger than ± 10 percent of the estimate).

- Hypotheses to adjudicate. A classic example is to statistically test whether one system is better than another under some input criteria, including the specification of the probabilities of incorrectly deciding the hypotheses. For example, test whether system A has a 70 percent or greater success rate with a 5 percent chance of incorrectly deciding it does not if the success rate is actually 70 percent, and a 10 percent chance of incorrectly deciding it does if the success rate is actually 60 percent.[3]

- Plausible existence of outcome. That is, see if, within specified ranges, an outcome is possible; e.g., is it possible Blue could lose X number of systems in a battle?

---

[3]With stochastic outcomes, this is the best we can do. That is, we can never be 100 percent certain, only with high probability.

- Identify extreme conditions—find maxima or minima within a region of model space.

- Assess sensitivity of model outcomes within a specified input region.

Once these precise statements are developed—and written down—the experiments can be designed to provide the information.

Case Selection Design

Case selection is the detailed specification of simulation inputs and replications. There are countless books on this subject and we will address only top-level issues and provide references for those desiring more detail. The design approach depends explicitly on the information that is needed from the experiments. Different classes of designs are appropriate for different types of experimental need. For large HITL ADS experiments there will be few samples feasible, and thus one should consider advanced designs (see below).

Estimation, Hypothesis Testing, and Traditional Designs

One method of adjudicating a hypothesis is to test a formal statistical hypothesis on variables within a model.[4] For example, one might want to test the hypothesis that in a given scenario system A has greater effectiveness than system B. In such a case, we need to estimate the effects of systems A and B, and perhaps some key interactions with other variables, with certain accuracy.[5] Unfortunately, a fact of life is that it takes at least one observation per item to uniquely estimate each main effect or interaction. Thus, the sample size required depends on what effects and interactions need to be estimated, as well as the required precision in the estimates. All of these factors must be weighed in selecting the class of design that will make the foundation for the run matrix. Few runs imply few effects can be estimated, which can be critical in deciding what ADS will investigate and what traditional methods will do.

---

[4]Testing the hypothesis within the model must still be related back to the analysis strategy. The fact that a hypothesis is accepted within a model does not necessarily guarantee that the hypothesis is valid in the real world.

[5]It is useful to distinguish the difference in estimating a variable (so-called main effect) and an interaction. A main effect is the average change in simulation outcome (say MOE) when a variable is set to a different value—with all other variables held constant. Two variables interact if the effect of one variable depends on the value of the other. If an interaction is present it typically makes little sense to talk about the main effects of the variables, for how they influence the outcomes depends on the levels (i.e., values) of the other variables. Similar reasoning holds for three-way and higher interactions.

If the number of variables that must be varied is relatively small, one can efficiently estimate and test for effects and interactions using traditional designs. By "traditional experimental design" we refer to those designs that are contained in the prominent texts and software on design of experiments (DOE). Good references on these include Box and Draper (1987), Cochran and Cox (1957), Kempthorne (1952), Montgomery (1991), and Taguchi (1976). For experiments where few variables are to be varied, full factorial designs are able to estimate all main effects and interactions. For experiments that are constructed to detect main effects and low-level interactions, fractional factorials are usually good to consider. A list of many complicated designs, where some variables and interactions are estimable at different levels, can be found in McLean and Anderson (1984). For estimating main effects only, main-effect screening designs,[6] such as Plackett-Burman designs and Latin Hypercubes,[7] should be considered. These are designed only to detect main effects, with the implicit assumption that variables with no main effects are likely not to have significant interactions—a risky assumption with many defense models. This difficulty can be mitigated if subsequent runs can be made.

The above designs have been studied and perfected for years. Today's designs provide the most information with the fewest samples. See Atkinson and Fedorov (1989) for more discussion and many references on the topic. These designs also tend to be robust to some of the assumptions, such as normality of error. In sum, traditional designs are recommended when comprehensively studying or testing hypotheses internal to a model on only a few variables.

Group Screening Designs

A wide variety of applications involve searching or screening multiple variables for outcomes that are believed quite rare. This is often the case in initial model exploration and sensitivity analysis. In such situations, one wants to screen as many variables as is possible with as few samples as is required. With HITL ADS, one cannot waste samples. Yet, we still need to ensure our results are robust to the numerous potentially confounding factors.

A class of experimental designs that do this efficiently is called group screening designs. For more on optimal group screening designs and the efficiencies

---

[6]For these designs, the interactions are confounded with the main effects, i.e., they cannot be distinguished from them. Therefore, a strong effect resulting from an interaction will, without additional investigation, be detected as a main effect.

[7]See McKay, Conover, and Beckman (1979) for the definition of Latin Hypercube designs.

obtained, as well as many more references on the subject, see Kotz and Johnson (1989).

Random Designs

It is often the case that there are uncertainties associated with a large number of input parameters, such as probabilities of kill, initial starting positions, decision thresholds, delay times, etc. If one only wants to investigate whether model outcomes are sensitive to these values, a simple approach is to randomly pick values of the variables, according to some specified distribution, for a few runs. In essence, we are independently shaking multiple input variables simultaneously to see if there are significant changes in model outcomes. While not optimal, it is relatively easy to do and will provide estimates of the sensitivity of the model to the numerous artificially fixed input variables.

Search Designs

Another class of design is relevant when one wants to identify extrema in a simulation—such as maxima or minima. These are the so-called search designs. They tend to be adaptive in that future case selection is determined by the outcomes of current runs. For simulations that are deterministic, there are several classic operations research algorithms (for example, see Bazarra and Shetty (1979)) that include hill climbing algorithms, such as steepest descent, which are appropriate for smooth unimodal response surfaces. For more complicated surfaces, such as multi-modal, one should consider simulated annealing and genetic algorithm approaches. "Response surface methodology" designs are appropriate for simulations that contain random components (see Box and Draper (1987)). Typically, these designs require more samples than will be available in the HITL ADS runs.

B. Community Actions to Facilitate Analysis

Update Analysis Guide

Members of the analysis community can contribute to the common good by building on this analysis guide and related documents. We consider this guide to be a living document, and we fully recognize our limited expertise in certain areas. Contributions from those with more expertise, and especially those with complementary backgrounds, will strengthen this guide enormously.

Additionally, the use of ADS for analytic purposes is in its infancy. No set of experts, or even the entire community, has all the answers—they have not been discovered yet. Indeed, it is safe to say we do not yet know all the problems. Periodic revisions to this guide will reflect the community's latest understanding of this new art.

Better Model Descriptions

One of the biggest problems for analysis with ADS is to understand the models and simulations that are used to build ADS federations. This is a problem with traditional stand-alone constructive analysis too, but for ADS it is exacerbated by the multiplicity of components and the geographical distribution of the expertise in these models and simulations.

The high-level architecture's (HLA's) intent of having Simulation Object Models (SOMs) available to provide descriptions of potential federates may mitigate this problem. However, we doubt that SOMs will successfully address all the needs of ADS analysts. The effort involved in building a SOM seems comparable to that of writing the highly detailed documentation required by now-defunct standards like 2167A. The fiscal and human resources required to produce such documentation can be prohibitive. More important, the resources required to *maintain* such documentation as the model/simulation is modified are unlikely, in our opinion, to be forthcoming for many of the potential federates. Undoubtedly, much of the routine effort involved in building and maintaining SOMs can be performed by automated tools operating in conjunction with computer-aided software engineering (CASE) systems, but these tools are typically poorly adapted for modification efforts and they will not be able to document critical information about underlying assumptions and simplifications.

For these reasons, we believe that SOMs will be useful primarily for "first-cut" culling and for interface construction when a federate is selected, but detailed interviews with the experts in each model will be necessary to truly assess suitability. This is a difficult undertaking, given geographic separation, and we think it likely that false starts with unsuitable federates will be quite common.

SOMs will nevertheless provide a valuable service, and the extent to which they will succeed in meeting the needs of analysts is unclear at this time. Therefore, recommendations on our part for specific steps to improve the ability of analysts to locate and assess potential federates for their ADS federations are premature, pending experience with SOMs. We do recommend, however, that the analytic community become closely involved in the HLA process, in order to maximize the capabilities of SOMs and other HLA tools to assist the analytic community.

Tool Enhancement and Sharing

Analysis with ADS will surely involve tasks that recur either many times within a single analysis or for many analytic efforts. These tasks will in many cases be subject to partial automation, and community efforts to build reusable tools that assist in the performance of tasks will be beneficial. The HLA should provide standards that assist in the development of reusable tools. Additionally, analysts need to be aware of potential synergies and develop their tools with anticipation of later reuse.

One issue that the analysis community should address head on is that of funding tools. It is often the case that building with reuse and extensibility in mind is not the best short-term solution. Accordingly, sponsors need to be educated about the longer-term benefits, so they will be willing to cooperatively fund the development of shared tools. Also, the analysis community needs to develop a vision/design for each tool, so that the tools that are developed can contribute parts to a reusable tool set.

As of this writing, it is premature to recommend specific tools, but it will be useful to enumerate several domains where automated tools should prove useful to analysts.

Configuration Management

Configuration management includes database and software configuration. The maintenance of consistent data across multiple sites and federates is difficult and error prone. We strongly feel that it is unwise to rely on humans to manually ensure data configuration consistency. Rather, we need automated tools for at least two steps. The first step is to have a centralized database (or at least a conceptual centralized database) that is provided with tools to generate data in the form needed by individual federates from the centralized data. When data change, the modifications can be made in one place, and then new local databases can be automatically generated from the revised master, assuring consistency.

This is no easy task, as evidenced by the complexity of AFSAA's MASTR project. The conversion of centralized data into a form usable by individual federates can involve complex "business rules," so the development of translators is far from a mechanical process. Additionally, these translators need to be maintained as the federates are updated. In fact, the MASTR project has retreated somewhat from complete centralization.

It is not sufficient to generate new federate databases automatically. It is also necessary to ensure that consistent versions of these databases are in fact used for exercises—another step where reliance on humans is unwise. We recommend the development of tools that can check local database versions at run initialization time, to guarantee consistency.

Analysis Objects and PDUs

The set of messages (or PDUs in the DIS context) needed to make a simulation function properly generally contains "what" information on simulation entities. Analysis also needs "why" information that is not in the minimal set of messages. For example, "what" information for a fighter aircraft includes when, if ever, the pilot launches an air-to-air missile, and what kind of weapon is selected. "Why" information in this example includes the reason for the selection of a particular weapon, and why the shot was (or was not) taken at a particular time. The need to provide "why" information has become well recognized, and various mechanisms to address the need for analysis information are becoming available. The HLA, for instance, permits federates to publish lists of data that they can make available, say to analysis databases, on demand. A DIS effort for Warbreaker, using "data probes," addressed a similar need for compatible exercise components.

This is an excellent start toward addressing the needs of analysts in this regard. For example, it provides a standardized means by which an analysis data logger can ask for and use information available from an entity. However, there is another side to the story: the need to easily provide for unanticipated data needs. In our experience, *every* analytic effort of any significant scope demands new data from the simulations it uses, a fact that implies that the set of data available for publication by a federate will inevitably be incomplete.

The point is that an attitude of "okay, let's just add the data to the public list" is not going to work very well. While adding the data is a conceptually easy thing to do, we are willing to bet that in actual simulations a number of steps for code changes, schema file changes, and the like will need to be made. In the case of HLA, there are also implied changes to the individual federate's SOM and the entire federation object model (FOM). All these changes are tedious and error prone. In our experience, it is fairly easy and (in the long run) highly advantageous to develop tools that can automatically make the necessary changes from a single input.

Accordingly, we strongly recommend that when the need to publish simulation data arises, either for the original development of a federate or its adaptation to HLA, that code generation and other appropriate automated tools be developed and used as early as possible.

Data Reduction

The reduction of ADS exercise data for analysis purposes is addressed in Section 8's subsection on data analysis. The point to be made here is that the tools for data consolidation and reduction have reuse potential within the community. Analysts developing these tools should strongly consider facilitating reuse.

Displays

Displays were discussed in Section 8's subsection on visualization tools. As in the case of data reduction, the HLA standards should present many reuse opportunities, resulting in better display capabilities than would be available if each user had to develop his own.

Collaborative Environments

This subsection is concerned with communication among the geographically distributed participants in an ADS exercise. A benefit to having a group working in close quarters is that sometimes one overhears a problem or issue being discussed and has useful information to contribute. It can work the other way too, where one might remember hearing a conversation that is relevant to one's current problem.

In principle, it ought to be possible to know who to go to get advice on each issue, but in practice this simply is not the case, particularly in a geographically distributed environment. A practical substitute for the group workplace is needed.

One idea, already in occasional use, is to conduct technical communication on a reflector, deliberately making it easy for others to "overhear" the discussion and contribute or learn. This is admittedly cumbersome when the individual you need to talk to is next door, and a simple reflector concept may not be ideal. However, there are real advantages that make it attractive even for local communications. Written communications avoid misunderstandings associated with verbal directives, and can provide an audit trail giving, for example, the underlying rationales for certain decisions. Devastating errors caused by changing old code and not being aware of modeling assumptions that depend on that code can be avoided. The result is usually unwanted behaviors and simulated phenomena, rather than an overt error.

Share Success and Failure Stories

There are a dearth of formal ADS for analysis publications. Indeed, the special *PHALANX* issue on "Advanced Distributed Simulation (ADS)/Distributed Interactive Simulation (DIS)," June 1995, contained no example ADS analysis efforts.

Some kind of repository of lessons learned would be helpful for analysts who may be faced with situations that others have previously encountered. It would be particularly useful for the next few years, when analysis with ADS is very much an art, not a science. Formal conferences, or sessions at MORS where "war stories" can be exchanged, can be useful. An Internet forum or (access-controlled?) newsgroup may be an excellent alternative means of disseminating this kind of information. If such a forum was linked to a searchable repository, it might become an extremely valuable resource. The repository idea would be a low-cost (low-maintenance) site where correspondence would be stored and could be searched using one of the automated search tools now available. The information in the repository should include tips and pitfalls, interesting phenomena that came to light at after action reviews, and information about reusable (or potentially reusable) tools that have been developed.

Virtual Red Opposing Forces

The vast majority of simulators are of friendly systems. While expensive, there are some analytic (and training) benefits that would accrue by having more virtual (simulator) opposing forces (OPFORs). Foremost among these benefits is that it would allow for more realistic and adaptive opposing forces in the simulations. It might also reduce some of the analytic bias that results from mixing HITL simulations playing Blue entities with constructive models representing Red entities. This is, of course, feasible only in larger virtual simulation sites.

Bibliography

Anderson, R., S. Bankes, P. Davis, H. Hall, and N. Shapiro, *Toward a Comprehensive Environment for Computer Modeling, Simulation, and Analysis*, RAND, N-3554-RC, 1993.

Atkinson, A., and V. Fedorov, "Optimum Design of Experiments," *Encyclopedia of Statistical Sciences: Supplement Volume*, Wiley, NY, pp. 107–114, 1989.

Bankes, S., "Exploratory Modeling for Policy Analysis," *Operations Research*, Vol. 41, No. 3, 1993, pp. 435–449.

Bazarra, M., and C. Shetty, *Nonlinear Programming: Theory and Algorithms*, Wiley, NY, 1979.

Beizer, B., *Software Testing Techniques*, Van Nostrand Reinhold, NY, 1990.

Boehm, B., *Software Engineering Economics*, Prentice-Hall, Englewood Cliffs, NJ, J-update 1981.

Box, G., and N. Draper, *Empirical Model Building and Response Surfaces*, Wiley, NY, 1987.

Brooks, A., S. Bankes, and B. Bennett, *Weapon Mix and Exploratory Analysis: A Case Study*, RAND, DB-216/2-AF, 1997.

Callero, M., C. Veit, E. Gritton, and R. Steeb, *Enhancing Weapon System Analysis: Issues and Procedures for Integrating a Research and Development Simulator with a Distributed Simulation Network*, RAND, MR-340-ARPA, 1994.

Cochran, W., *Sampling Techniques*, Wiley, NY, 1977.

Cochran, W. G., and G. M. Cox, *Experimental Designs*, 2nd Edition, Wiley, NY, 1957.

Cox, A., E. Luiffjf, R. Van Kampen, and R. Ripley, "Time Synchronization Experiments," *Summary Report, 14th DIS Workshop*, Institute for Simulation and Training, March 1996.

Davis, P., *An Introduction to Variable-Resolution Modeling and Cross-Resolution Model Connection*, RAND, R-4252-DARPA, 1993.

Davis, P., "Distributed Interactive Simulation in the Evolution of DoD Warfare Modeling and Simulation," *Proceedings of the IEEE*, Vol. 83(8), August 1995, pp. 1138–1155.

Davis, P., *Generalizing Concepts and Methods of Verification, Validation, and Accreditation (VV&A) for Military Simulations*, RAND, R-4249-ACQ, 1992.

Davis, P., "Modeling of Soft Factors in the RAND Strategy Assessment System (RSAS)," Military Operations Research Society, *Mini-Symposium Proceedings: Human Behavior and Performance as Essential Ingredients in Realistic Modeling of Combat-MORIMOC II*, Alexandria, VA, 1989; also RAND, P-7538, 1989.

Defense Modeling and Simulation Office (DMSO), *Verification, Validation, and Accreditation (VV&A) Recommended Practices Guide*, 1996. See the web site at "http:/www.dmso.mil/docslib/" for updated versions.

Defense Modeling and Simulation Office (DMSO), *Modeling and Simulation (M&S) Master Plan*, Office of the Under Secretary of Defense for Acquisition and Technology, DoD 5000.59-P, Washington, D.C., 1995.

Defense Science Board, Task Force on Simulation, Readiness, and Prototyping, *Impact of Advanced Distributed Simulation on Readiness, Training and Prototyping*, January 1993.

Department of the Air Force, *A New Vector*, June 1995.

Department of the Army, *Distributed Interactive Simulation (DIS) Master Plan*, September 1994.

Dewar, J., S. Bankes, J. Hodges, T. Lucas, D. Saunders-Newton, and P. Vye, *Credible Uses of the Distributed Interactive Simulation System*, RAND, MR-607-A, 1996.

Don, B., J. Friel, and T. Herbert, *Ground Commander's Close Support Needs and Preferred System Characteristics*, RAND (forthcoming).

Dupuy, T., *Understanding War*, Paragon House Publishers, NY, 1987.

Efron, B., and R. Tibshirani, *An Introduction to the Bootstrap*, Chapman & Hall, NY, 1993.

Fisher, R., *The Design of Experiments*, 4th ed., Oliver and Boyd, Edinburgh, 1947.

Foster, L., and P. Feldman, "Defining a Fair Fight," *14th DIS Workshop*, Volume I, Position Papers, March 1996, pp. 147–153.

Garrett, R., "Architectural Design Considerations," *PHALANX*, Vol. 29, No. 2, June 1996.

Harmon, S., "Interoperability Between Distributed Simulations I: Interacting Models of Physical Processes," *15th DIS Workshop*, Volume I, Position Papers, September 1996, pp. 275–285.

Hillestad, R., J. Owens, and D. Blumenthal, *Experiments in Variable Resolution Combat Modeling*, RAND, N-3631-DARPA, 1993.

Hodges, J. S., "Six (or so) Things You Can Do with a Bad Model," *Operations Research*, Vol. 39, No. 3, May-June 1991, pp. 355–365.

Hodges, J. S., and J. A. Dewar, *Is It You or Your Model Talking? A Framework for Model Validation*, RAND, R-4114-AF/A/OSD, 1992.

Institute for Simulation and Training, *Exercise Management & Feedback (EMF)—Rationale*, IST-CR-96-16, October 1996. Available on FTP: http://ftp.sc.ist.ucf.edu/SISO/dis/library/emfrati.doc

Institute for Simulation and Training, *Analysis & Engineering (A&E) DIS ++ Application Guidelines*, IST-TR-96-35, September 1996. Copies available through DMSTTIAC Service Center (Phone: 407-249-4712, FAX: 407-658-5059).

Institute for Simulation and Training, DIS Steering Committee, *Draft Standard for Distributed Interactive Simulation: Exercise Management & Feedback*, Document Set, March 1995.

Institute for Simulation and Training, DIS Steering Committee, *The DIS Vision: A Map to the Future of Distributed Simulation*, May 1994.

Jones, Anita, "ADS for Analysis," *PHALANX*, Vol. 29, No. 2, June 1996.

Kempthorne, O., *The Design and Analysis of Experiments*, Wiley, NY, 1952.

Kerchner, B., J. Friel, and T. Lucas, *Understanding the Air Force's Capability to Effectively Apply Advanced Distributed Simulation for Analysis*, RAND, MR-744-AF, 1996.

Kotz, S., and N. Johnson, "Dorfman-Type Screening Procedures," *Encyclopedia of Statistical Sciences: Supplement Volume*, Wiley, NY, 1989, pp. 50–53.

Law, A., and D. Kelton, *Simulation Modeling and Analysis*, McGraw-Hill, NY, 1991.

Lucas, T., S. Bankes, and P. Vye, *Improving the Analytical Contribution of Advanced Warfighting Experiments*, RAND, DB-207-A (forthcoming).

Marshall, C., and R. Garrett, "Simulation for C4ISR," *PHALANX*, Vol. 29, No. 1, March 1996.

Matsumura, J., R., Steeb, T. Herbert, M. Lees, S. Eisenhard, and A. Stich, *Analytic Support to the Defense Science Board : Tactics and Technology for 21st Century Military Superiority*, RAND, DB-198-A, 1997.

McKay, M., W. Conover, and R. Beckman, "A Comparison of Three Methods for Selecting Values of Input Variables in the Analysis of Output from a Computer Code," *Technometrics*, Vol. 21, No. 2, May 1979.

McLean, R., and V. Anderson, *Applied Factorial and Fractional Designs*, Marcel Dekker, NY, 1984.

Montgomery, D. C., *Design and Analysis of Experiments*, Wiley, NY, 1991.

MORS Workshop, "Advanced Distributed Simulation for Analysis (ADSA '96)," briefing, Williamsburg, VA, January 30–February 1, 1996.

Myjak, M., "DIS Time," *Summary Report, 14th DIS Workshop*, Institute for Simulation and Training, March 1996.

Pace, D., "An Aspect of Simulation Cost," PHALANX, Vol. 30, No. 1, March 1997.

Palmore, J., Editor, "Special Issue—Verification, Validation and Accreditation (VV&A) of Models and Simulations," PHALANX, Vol. 30, No, 2, June 1997.

Quade, E., *Analysis for Public Decisions*, North-Holland, NY, 1989.

Quality Research, *A Methodology Handbook for Verification, Validation, and Accreditation (VV&A) of Distributed Interactive Simulations (DIS)*, January 1995.

Redman, J., *PHALANX*, Vol. 28, No. 2, June 1995.

Russo, G., "Attaining Interoperability Between ModSAF and JANUS," *15th DIS Workshop, Volume I position papers*, September 1996, pp. 171–180.

Sanders, P., and R. Miller, "Model Verification, Validation, and Accreditation (VV&A) Common Ground Within the M&S Community," *PHALANX*, Vol. 29, No. 3, September 1996.

Smith, G., and K. Winkler, "Security Policies and Mechanisms for M&S Repositories," *Summary Report, 14th DIS Workshop*, Institute for Simulation and Training, March 1996.

Steeb, R., J. Matsumura, T. Covington, S. Eisenhard, and L. Melody, *Rapid Force Projection Technologies: A Quick-Look Analysis of Advanced Light Indirect Fire Systems*, RAND, DB-169-A/OSD, 1996.

Stender, J. F., "Simulation Management for Large Exercises," *Summary Report, 14th DIS Workshop*, Institute for Simulation and Training, March 1996.

Taguchi, G., *An Introduction to Quality Control*, Central Quality Control Association, Nagoya, Japan, 1976.

Tiernan, T., K. Boner, C. Keune, and D. Coppock, *Synthetic Theater of War-Europe (STOW-E): Technical Analysis*, Naval Command, Control and Ocean Surveillance Center, May 1995.

Wald, A., *Sequential Analysis*, Wiley, NY, 1947.

L. Wall, T. Christiansen, and R. Schwartz, *Programming for PERL*, 2nd Edition, O'Reilly & Associates, Cambridge, MA, 1996.